GENETIC DESTINY

Published for the Society for the Psychological Study of Social Issues

GENETIC DESTINY

Race
as a
Scientific
and Social
Controversy

*edited
by
Ethel Tobach
and
Harold M. Proshansky*

PHILLIPS MEMORIAL
LIBRARY
PROVIDENCE COLLEGE

AMS PRESS
New York

BF
76.5
G46

Copyright © 1976 AMS Press, Inc.

All rights reserved. Published in the United States by AMS Press, Inc., 56 East 13th Street, New York, New York 10003

Library of Congress Cataloging in Publication Data
Main entry under title:

Genetic destiny.

    Includes index.
    1. Psychological research—United States. 2. Intelligence tests. 3. Race discrimination. 4. Jensen, Arthur Robert. 5. Nature and nurture. I. Tobach, Ethel, 1921-   II. Proshansky, Harold M., 1920-
BF76.5.G46   155.2'34   765964
ISBN 0-404-10130-5

Manufactured in the United States of America

CONTENTS

July, 1972	3	Comment
December 6, 1972	8	A Letter to the Fifty Signers
April 3, 1973	11	An Inquiry
Harold M. Proshansky	13	Social Response and the Politics of Scientific Controversy
Abraham Edel	31	The Scientist and His Findings: Some Problems in Scientific Responsibility
Robert Cancro	49	Race, Reification, and Responsibility
David Layzer	59	Behavioral Science and Society: The Nature-Nurture Controversy as a Paradigm
Harland Padfield	80	Social Marginalization: Toward a General Theory of Inequality

Allan Chase	99	False Correlations Equal Real Deaths: The Great Pellagra Cover-up, 1914-1933
Ernest Drucker *Victor W. Sidel*	113	Scientific Freedom: Sacred Principle or Secular Politics?
Gerard Piel	123	". . . Ye May Be Mistaken"
James C. King	134	Padfield-Herrnstein and Layzer-Jensen
Evelyn A. Mauss	137	The Science Information Movement: Non-Partisan but Not Neutral
Ethel Tobach	142	Behavioral Science and Genetic Destiny: Implications for Education, Therapy, and Behavioral Research
	159	Index

ACKNOWLEDGMENTS

The editors and contributors wish to express their appreciation to the New York Foundation for its support in the preparation of this book. They also wish to thank The Society for the Psychological Study of Social Issues for its encouragement. The royalties from the sales of this book will go to SPSSI.

Finally, it should be noted that the authors' views and beliefs expressed in this book do not reflect the official policies or programs of either the New York Foundation or SPSSI.

GENETIC DESTINY

Race as a Scientific and Social Controversy

COMMENT

July, 1972

BEHAVIOR AND HEREDITY

The posthumous Thorndike Award article by Burt (1972) draws psychological attention again to the great influence played by heredity in important human behaviors.(1) Recently, to emphasize such influence has required considerable courage, for it has brought psychologists and other scientists under extreme personal and professional abuse at Harvard, Berkeley, Stanford, Connecticut, Illinois, and elsewhere. Yet such influences are well documented. To assert their importance and validity, and to call for free and unencumbered research, the 50 scientists listed below have signed the following document, and submit it to the APA:

Background: The history of civilization shows many periods when scientific research or teaching was censured, punished, or suppressed for nonscientific reasons, usually for seeming to contradict some religious or political belief. Well-known scientist victims include: Galileo, in orthodox Italy; Darwin, in Victorian England; Einstein, in Hitler's Germany; and Mendelian biologists, in Stalin's Russia.

Today, a similar suppression, censure, punishment, and defamation are being applied against scientists who emphasize the role of heredity in human behavior. Published positions are often misquoted and misrepresented; emotional appeals replace scientific reasoning; arguments are directed against the man rather than against the evidence (e.g., a scientist is called "fascist," and his arguments are ignored).

A large number of attacks come from nonscientists, or even antiscientists, among the political militants on campus. Other attackers include academics committed to environmentalism in their explanation of almost all human differences. And a large number of scientists, who have studied the evidence and are persuaded of the great role played by heredity in human behavior, are silent, neither expressing their beliefs clearly in public, nor rallying strongly to the defense of their more outspoken colleagues.

The results are seen in the present academy: it is virtually heresy to express a hereditarian view, or to recommend further study of the biological bases of behavior. A kind of orthodox environmentalism dominates the liberal academy, and strongly inhibits teachers, researchers, and scholars from turning to biological explanations or efforts.

Resolution: Now, therefore, we the undersigned scientists from a variety of fields, declare the following beliefs and principles:

1. We have investigated much evidence concerning the possible role of inheritance in human abilities and behaviors, and we believe such hereditary influences are very strong.

2. We wish strongly to encourage research into the biological hereditary bases of behavior, as a major complement to the environmental efforts at explanation.

3. We strongly defend the right, and emphasize the scholarly duty, of the teacher to discuss hereditary influences on behavior, in appropriate settings and with responsible scholarship.

4. We deplore the evasion of hereditary reasoning in current textbooks, and the failure to give responsible weight to heredity in disciplines such as sociology, social psychology, social anthropology, educational psychology, psychological measurement, and many others.

5. We call upon liberal academics—upon faculty senates, upon professional and learned societies, upon the American Association of University Professors, upon the American Civil Liberties Union, upon the University Centers for Rational Alternatives, upon presidents and boards of trustees, upon departments of science, and upon the editors of scholarly journals—to insist upon the openness of social science to the well-grounded claims of biobehavioral reasoning, and to protect vigilantly any qualified faculty members who responsibly teach, research, or publish concerning such reasoning.

We so urge because as scientists we believe that human problems may best be remedied by increased human knowledge, and that such increases in knowledge lead much more probably to the enhancement of human happiness than to the opposite.

Signed:

Jack A. Adams
Professor of Psychology
University of Illinois

Dorothy C. Adkins
Professor/Researcher in Education
University of Illinois

Andrew R. Baggaley
Professor of Psychology
University of Pennsylvania

Irwin A. Berg
Professor of Psychology and
    Dean of Arts & Sciences
Louisiana State University

Edgar F. Borgatta
Professor of Sociology
Queens, College, New York

Robert Cancro, MD
Professor of Psychiatry
University of Connecticut

Raymond B. Cattell
Distinguished Research Professor
    of Psychology
University of Illinois

Francis H.C. Crick
Nobel Laureate
Medical Research Council
    Laboratory of Molecular Biology
Cambridge University

C.D. Darlington, FRS
Sherardian Professor of Botany
Oxford University

Robert H. David
Professor of Psychology and
    Assistant Provost
Michigan State University

M. Ray Denny
Professor of Psychology
Michigan State University

Otis Dudley Duncan
Professor of Sociology
University of Michigan

Bruce K. Eckland
Professor of Sociology
University of North Carolina

Charles W. Eriksen
Professor of Psychology
University of Illinois

Hans J. Eysenck
Professor of Psychology
    Institute of Psychiatry
University of London

Eric F. Gardner
Slocum Professor & Chairman
Education and Psychology
Syracuse University

Benson E. Ginsburg
Professor & Head,
    Biobehavioral Sciences
University of Connecticut

Garrett Hardin
Professor of Human Ecology
University of California,
    Santa Barbara

Harry S. Harlow
Professor of Psychology
University of Wisconsin

Richard Herrnstein
Professor & Chairman of Psychology
Harvard University

Lloyd G. Humphreys(2)
Professor of Psychology
University of Illinois

Dwight J. Ingle
Professor and Chairman
    of Physiology
University of Chicago

Arthur R. Jensen
Professor of Educational Psychology
University of California, Berkeley

Ronald C. Johnson
Professor & Chairman of Psychology
University of Hawaii

Henry F. Kaiser
Professor of Education
University of California, Berkeley

E. Lowell Kelly
Professor of Psychology &
  Director, Institute of Human
  Adjustment
University of Michigan

John C. Kendrew
Nobel Laureate
MRC Laboratory of Molecular Biology
Cambridge, England

Fred N. Kerlinger(2)
Professor of Educational Psychology
New York University

William S. Laughlin
Professor of Anthropology &
  Biobehavioral Sciences
University of Connecticut

Donald B. Lindsley
Professor of Psychology
University of California, Los Angeles

Quinn McNemar
Emeritus Professor of Psychology,
  Education, and Statistics
Stanford University

Paul E. Meehl
Regents Professor of Psychology and
  Adjunct Professor of Law
University of Minnesota

Jacques Monod
Nobel Laureate
Professor, Institute Pasteur
College de France

John H. Northrup
Nobel Laureate
Professor Emeritus of Biochemistry
University of California and
  Rockefeller University

Lawrence I. O'Kelly
Professor and Chairman of Psychology
Michigan State University

Ellis Batten Page
Professor of Educational Psychology
University of Connecticut

B.A. Rasmusen
Professor of Animal Genetics
University of Illinois

Anne Roe
Professor Emerita, Harvard
  University & Lecturer in Psychology
University of Arizona

David Rosenthal
Research Psychologist and Chief
  of Laboratories
National Institute of Mental Health

David G. Ryans
Professor & Director Educational
  R & D Center
University of Hawaii

Eliot Slater, MD
Professor of Psychiatry and Editor
British Journal of Psychiatry
University of London

H. Fairfield Smith
Professor of Statistics
University of Connecticut

S.S. Stevens
Professor of Psychophysics
Harvard University

William R. Thompson
Professor of Psychology
Queens University, Canada

Robert L. Thorndike
Professor of Psychology and Education
Teachers College
Columbia University

Frederick C. Thorne, MD
Editor,
Journal of Clinical Psychology
Brandon, Vermont

Philip E. Vernon
Professor of Educational Psychology
University of Calgary, Alberta

David Wechsler
Professor of Psychology
N.Y.U. College of Medicine

Morton W. Weir
Professor of Psychology and
  Vice-Chancellor
University of Illinois

David Zeaman
Professor of Psychology and NIMH
  Career Research Fellow
University of Connecticut

NOTES

(1) Burt, C. Inheritance of general intelligence. *American Psychologist*, 1972, 27, 175-190.

(2) In Item 1, preferred "substantial" or "important" to the wording "very strong."

A LETTER TO THE FIFTY SIGNERS OF "COMMENT"

December 6, 1972

We are writing about the "Comment" appearing over your signature in the July 1972 *American Psychologist.* We have been constituted by the Council of SPSSI as a Commission to examine for purposes of later publication the social meaning of that "Comment." In order to understand the significance of that "Comment," we are seeking information about some of its statements, as shown below.

1. We would also deplore the ". . . personal and professional abuse at Harvard, Berkeley, Stanford, Connecticut, Illinois and elsewhere. . . ." as distinguished from appropriate professional criticism and would like to know of any specific personal experiences you may have had, as well as of any such incidents that may have occurred at your institution, or any other. We would also appreciate any information you may have about the events referred to at the five institutions mentioned in the "Comment."

2. We would like to have information also about the "suppression, censure, punishment, and defamation . . ." of ". . . scientists who emphasize . . . the role of heredity . . . in human behavior."

3. Which "academics committed to environmentalism in their explanation of almost all human differences" have been "attackers" of whom?

4. Where is it "heresy to express" which "hereditarian view, or to recommend further study of the biological bases of behavior?" Who and in what way are which "teachers, researchers and scholars" . . . "strongly inhibit(ed)" from "turning to" which "biological explanations of efforts?"

5. We are most interested in any instances in which a "hereditarian"

viewpoint has been the basis on which a scientist lost a position in an academic institution, was prevented from teaching, was prevented from doing research, or was not allowed to publish in appropriate scientific journals.

We defend the rights of scientists to engage in such activities as analysis, criticism, exchange of information and interpretation of controversial issues in science. We also uphold the right of scientists to present their ideas about controversial subjects to the public at large and to engage in free public discussion. Finally, we uphold the public's right to respond to the presentation of these issues.

For these reasons, we are interested in learning about the events responsible for statements appearing in the "Comment" section of the *American Psychologist*. We hope you will respond as quickly as possible to our request. As you can see from the accompanying letter sent to the *American Psychologist*, we wish to present the results of this enquiry in that journal. We also plan to organize a symposium on these issues at the Montreal meetings of the American Psychological Association.

Signed:

Harold M. Proshansky, Chair
SPSSI Commission on the Renewed Assault on Equality
Acting President and Provost
The Graduate School and University Center
The City University of New York

Cynthia Deutsch
Professor
Institute for Developmental Studies
New York University

Martin Deutsch
Professor and Director
Institute for Developmental Studies
New York University

Morton Deutsch
Professor
Department of Psychology
Teachers College
Columbia University

Lee Ehrman
Associate Professor of Natural Sciences
State University of New York at Purchase

Howard Gruber
Professor
Institute for Cognitive Studies
Rutgers University

Jerry Hirsch
Professor
Department of Psychology and Zoology
University of Illinois

James C. King
Professor
Department of Microbiology
New York University College of Medicine

Thomas F. Pettigrew
Professor
Department of Social Relations
Harvard University

Ethel Tobach
Curator
Department of Animal Behavior
American Museum of Natural History

Curtis Williams
Dean of Natural Sciences
State University of New York at Purchase

AN INQUIRY

April 3, 1973

In response to a "Comment" published in the July, 1972 *American Psychologist* (p. 660), SPSSI Council at its August, 1972 meeting unanimously passed the resolution that a SPSSI Commission be constituted in New York on the new assault on equality with a mandate to try to do the following: 1. Respond to the letter, "Behavior and Heredity," in the July *American Psychologist* signed by fifty psychologists in a point by point critique; 2. Prepare a SPSSI policy statement on the issues for Council to consider in February for submission to the APA and a wider audience; 3. Prepare and create a major conference, within the next year, inviting social scientists, minority groups, people within the media, to participate in an effort to build communication among all groups on this issue; 4. Urge people to establish research concerned with anti-egalitarian trends in society immediately for possible use by the Conference.

Subsequently, the Commission wrote to the fifty signers of that "Comment" and to the *American Psychologist*, asking the following five questions:

1. We would also deplore the ". . . personal and professional abuse at Harvard, Berkeley, Stanford, Connecticut, Illinois and elsewhere. . . ." as distinguished from appropriate professional criticism and would like to know of any specific personal experiences you may have had, as well as of any such incidents that may have occurred at your institution, or any other. We would also appreciate any information you may have about the events referred to at the five institutions mentioned in the "Comment."

2. We would like to have information also about the "suppression, censure,

punishment, and defamation . . ." of ". . . scientists who emphasize . . . the role of heredity . . . in human behavior."

3. Which "academics committed to environmentalism in their explanation of almost all human differences" have been "attackers" of whom?

4. Where is it "heresy to express" which "hereditarian view, or to recommend further study of the biological bases of behavior?" Who and in what way are which "teachers, researchers and scholars" . . . "strongly inhibit(ed)" from "turning to" which "biological explanations or efforts?"

5. We are most interested in any instances in which a "hereditarian" viewpoint has been the basis on which a scientist lost a position in an academic institution, was prevented from teaching, was prevented from doing research, or was not allowed to publish in appropriate scientific journals.

Of the twenty-six respondents several criticized the name of the Commission as indicating a lack of objectivity, and questioned the impartiality of some of the Commission members. The Commission, wishing to avoid diversionary controversy and to concentrate on its originally stated objective of ascertaining the facts, changed its name to "Fact Finding Commission on the Suppression of Academic and Scientific Freedom of Hereditarian and Behavioral Researchers and Teachers." A letter was then sent to the fifty signers informing them of this action. The change did not produce any substantial increase in information.

A group of the respondents informed us that they have requested the Board of Scientific Affairs to carry out an "objective and unbiased study of the . . . factors" leading to the resolution. It is important to note that one member of the SPSSI Commission had tried unsuccessfully before the publication of the Comment in July to get the Committee on Social and Ethical Responsibility (now a Board with the same name) to determine whether academic freedom had been violated as charged in the "Comment" and to take action to protect any offended individuals.

Because we are committed to universal intellectual and academic freedom, and we believe the issues raised by the "Comment" are of immense importance we have undertaken the following actions:

1. We have concluded that it would be constructive to make the material received available to all who wish to examine and evaluate it, rather than attempt to do so ourselves or choose a panel of "elder statesmen" for this purpose, as we had previously suggested. We are also sending all the material to the Board of Scientific Affairs. Of course the signers will be asked for clearance of their letters first. Those who wish to obtain copies of the material should send $1.00 to cover mailing costs to Mrs. Caroline Weichlein etc.

2. We are requesting SPSSI to devote an issue of the *Journal of Social Issues* to original documentation, articles and comment on various aspects of the publication of the "Comment" and subsequent events. The fifty signers will be invited to participate in this issue.

3. Symposia will be organized at future APA conventions to discuss two major aspects of the issues: Academic Freedom in its Broadest Sense and Genetics-Environment-and Behavior.

SOCIAL RESPONSIBILITY AND THE POLITICS OF SCIENTIFIC CONTROVERSY

Harold M. Proshansky

The "story" to be told in this book began with a letter which appeared under the title, "Comment" in *The American Psychologist* about four years ago; in it was a resolution signed by fifty well known and distinguished behavioral and biological scientists (1972). "Comment" is republished at the beginning of this book, because it is in fact the nexus of the story we are about to tell and which took place over the two years following the publication of "Comment."

SOME BACKGROUND

If the publication of "Comment" was the first of the events we will relate below, then in a way it can be said that the publication of this volume—or certainly the publication of this chapter represents its end. Before looking at these events and the "piece of research" that grew out of them, it is important to provide a context within which to understand the meaning and significance of these events and the "research" that was done.

At the heart of the matter is the scientific conflict over the relative significance of hereditary and environmental determinants in explaining complex human characteristics and dispositions, particularly the intelligence of racial groups. Although this conflict has waxed and waned over many decades, it crystallized into a major dispute in America and abroad with the findings and interpretations presented by Arthur Jensen (1969), and subsequently in the continuing debate between Jensen and other behavioral scientists.

It should be evident from the title of this paper, that it is not the scientific issue or theoretical controversy itself that is under consideration. What are of concern are the "politics" of this controversy and its implications for the broader question of science and its social responsibility. By the politics of scientific controversy we mean the process by which the conflict over the theoretical issue itself reveals and reinforces more fundamental value or ideological differences underlying the controversy; these revealed the differences in turn serving to influence attitudes, behavior, and public policy orientations of those in the scientific community as well as those outside of it who hold public positions of authority and responsibility.

What cannot be stressed too strongly, however, is the fact that the political process and not the actual research endeavor becomes the evaluative context for judging the validity of the positions involved in the theoretical controversy as they are revealed in differences over more specific theoretical or empirical questions. Like any bitterly contested political campaign for public office, it is the public pronouncements, the debates in public, the side that attracts the greatest number, and indeed even the energy and ingenuity of either side in stating their case, that becomes the measure of which view will prevail as the "valid" scientific one in the case. To give but one example: to the extent that Jensen experienced personal abuse at the hands of extreme groups when he appeared at public meetings, it became difficult at times to distinguish defenses of Jensen's right to be heard from defenses of the value and significance of his research findings.

Having defined the "politics of scientific controversy," it is important to provide a more detailed framework as the conflict involved concerned the intelligence of racial groups. As we have noted, the heredity-environment issue in the development of human intelligence is by no means new. Yet, it should be clear the scientific controversy did not begin *in vacuo*, and then have effects on public policy. Indeed the reverse was true. The strong need to measure human learning potential in order to relate pupil capacities to level of primary education, and similarly the considerable task of training and therefore distinguishing levels of learning ability among army personnel during World War I, resulted in the development of theory and data relevant to the nature of human intelligence, how it could be measured, and the external criteria to be used in testing the validity of such measurements. As Allan Chase makes clear in this volume, it is not a value-free science that provides the "blueprint" for the way a society structures (or should structure) the functions, relationships, and rewards and goals of the individuals and groups of which it is comprised. On the contrary, it is the social, economic and political strands of a society that are woven into the institutional fabric of science, thereby influencing the selection of problems for study, the methods to be used in their study, and even the kinds of solutions that will be acceptable. Science is no more value-free in its defined purposes, meanings, and methods than any other institutional structure.

If we accept this assumption—and remember our focus is on the *politics* of the hereditary-environment controversy—then our historical framework of this controversy can be more fully detailed. Although by no means a perfect distinction, by and large those who supported the hereditary-genetic view explanation of intelligence were expressing a fundamental difference in approach to psychology as a science compared to those who saw the influence of the sociocultural setting and human experiences as primary factors in explaining group differences in intelligence. For those ready to accept the meager and questionable research findings in support of the role of hereditary factors in intelligence, psychology as a science was clearly one of the physical and natural sciences. Although obviously behind in its development as a science, the philosophical and methodological path for psychology to take its rightful place in the hierarchy of other sciences was clear enough. It had to be experimental and quantitative; it had to make use of the laboratory in its research orientation; above all it had to seek the underlying principles that would explain all of human behavior. Understanding human intelligence or other complex manifestations of human behavior—attitudes, values, neurotic behavior, violence, aggression and interpersonal and group relationships—was simple enough from a theoretical viewpoint: since much evidence of continuity in levels of biological organization and functions exists, it could be safely assumed that systematic research at any one of these levels could reveal for all of these a single set of unifying principles of perception, learning, motivation. Once these principles were established all the rest would fall into place. The complexities of human behavior were more apparent than real. The view then was reductionistic, "hard-nosed," and completely intolerant of approaches and concepts that conceived of the person as a cognitive, goal-directed social organism whose understanding required new levels of analysis, because the psychological, social and cultural systems that defined a person's existence revealed their influence in the unique, emergent properties of human behavior and experience. (Schneirla, 1971; Tobach, 1972)

Given this sharp methodological schism in scientific psychology—particularly during the period of the 1930's through the 1950's—other consequences unfolded in very predictable ways. Thus, those in the "hard-nosed" areas of specialization not only assumed the highest status as scientific psychologists (after all they were in fact simply emulating the postures and approaches of the other scientists who had established the credibility of physics, chemistry, or biology, but also took the lead in establishing the academic-research character of psychology as a college or university discipline. Departments of psychology were either almost exclusively experimental with little if any representation on the faculty given to social, personality, clinical and development psychology, except if the approach in these fields were consistent with the physical-natural scientist model. Or, if due to the particular historical development of the department, these fields were given reasonable representation in the curriculum, in most instances the courses and faculty involved were accorded a secondary if not inferior status. As for an applied field like clinical psychology, which did not evolve out of experimental psychology as an area of specialization (in the way, for example, that the field of "human factors" did), its "very poor relation" status among the other fields of specialization in a department needs no retelling.

Of course during this period the leadership, policies, and concerns of the American Psychological Association also reflected this far greater status and influence of the physical-natural scientist model. What must be emphasized as essential in this model was—and still is—its dedication to a highly operational-empirical approach in which fundamental principles in the form of precise quantitative relationships concerning human behavior and experience can be derived from systematic, highly controlled investigations of the psychological, physical, and biological processes that characterize all such behavior and experience at the more complex levels of the phyletic scale. During this thirty year period of growth of psychology both as science and a profession, the then so called "soft" specializations in the field, e.g. social, personality, also revealed the influence of the methodological schism in the field as a whole.

In effect psychologists in these fields simply applied in both their theory and research the physical-natural scientist paradigm described above to the more complex human problems that concerned them. Others of course rejected this model either in part or totally, focusing on approaches which made use of concepts, theoretical structures, and relevant methods and techniques that emerged from and maintained the integrity of these complex problems. The controversy over the role of hereditary-environmental factors in intelligence was to be found within these specialization areas, e.g. social, developmental, personality, as well as in the field of psychology as a whole.

In the 1950's and 1960's psychology as a scientific field grew quickly; new doctoral training programs, new specializations, consultations in community and governmental problems or issues, research projects, and publications were established. It was the increased numbers of psychologists in the soft fields, the availability of funding for "human problems," and the possibility of achieving the "great society" through intervention in major social problems, that curtailed if not equalized the influence of *the experimental scientist* model. However, one negative factor mitigated this influence: this model had failed to deliver at least at the level it promised during the 1940's and 1950's. Whether in learning, motivation, emotion, or perception, the promised "underlying principles" failed to emerge; controversies within the respective camps of the experimentalists in these fields were no less evident. Yet, the *complexity of the human species* seemed to taunt and defy all theoretical and methodological orientations, e.g. experimentalists, holists, psychoanalytic theorists, etc., if for no other reason than all these orientations tended to search for a *simple sovereign theory* that would provide one understanding of human behavior and experience.

Thus, the social, clinical, personality, developmental, and other psychologists, failed to solve social problems, whether they worked in a laboratory for small group research or attitude change or worked in a "head start school," a mental alth community research unit, or a school desegregation program.

the late 1960's and early 1970's the various specializations for fields of psych 'ogy—especially within the organizational framework of the American Psycholc. 'cal Association and its regional associations—were in serious trouble, and that trouble continues at the present time. In terms of questions of scientific and professional policies, the role of psychology in and out of the university, the commitment of psychologists to social and political issues, the development of psychology as a science, and other fundamental dilemmas, the field is splintered into numerous groups who at one moment are ready to go their separate ways and at others to reorganize into a loose confederation to allow each sub-discipline or specialization to realize its own value commitments.

Psychology can be regarded as a scientific field of inquiry in a malaise; the field of psychology is not quite sure where it is heading, what it is supposed to do, nor how to resolve its problems. To regard the American Psychological Association as a collection of groups that simply reflect differences in substantive interests, is to belie the fundamental conflicts of purpose, means, and goals that now divide academics from applied researchers, professional practitioners from basic scientists, and social issue oriented psychologists from scientific purists. The reasons for this "breakdown" in a growing and thriving field that once had its purpose and destiny firmly established are simple enough to enumerate: it grew and diversified both as a scientific discipline and as a practitioner profession at an extraordinary rate. Scientifically and professionally, it failed to deliver what it had promised. It assumed roles and purposes in the community for which it was neither prepared or ready to adapt to. It found itself confronted with moral and ethical issues in its research and application that it largely ignored for more than two decades. Although psychologists were committed to the improvement of mankind in a democratic society—as well as to the development of psychology as a basic science—the accomplishments of psychology in terms of the theories it formulated, the research it did, and the actual opportunities it gave to all ethnic minority groups and women were few and disappointing.

THE RESOLUTION AND WHAT HAPPENED

The resolution within "Comment" in the *American Psychologist* was widely read by psychologists, as this journal is the official publication of the American Psychological Association. It is sent each month to every one of the Association's forty thousand members. That it appeared in the July issue increased the possibility for some kind of organized reaction to be generated relatively quickly, because the APA holds its annual convention at the end of summer recess at the very end of August. At the 1972 convention held in Honolulu, a number of members of Division 9 of the APA, also known as the Society for the Psychological Study of Social Issues (SPSSI), raised the question of the meaning of the resolution at the SPSSI Council meetings held a few days before the official opening of the APA convention.

This division of the APA is perhaps one of the oldest groups in the Association. It was formed in 1936 by a small number of psychologists primarily from the fields of social-personality and developmental psychology. Its emergence during the Great Depression and during the beginning of war in

Europe was no accident. The critical value that formed the group and maintains it today was the belief that psychologists and other behavioral scientists can play a role in resolving major social problems and issues by means of systematic research and the development of systematic, scientific knowledge. Consequently, the major human problems of poverty, unemployment, ethnic prejudice, intergroup conflict, war and peace, and religious conflict were and are the focus of the concern of SPSSI members.

SPSSI's membership was and is open to other behavioral scientists and social practitioners, such as social workers and school administrators. Although there was and continues to be differences among the members on how problem-oriented the organization should be, and moreover, on how much *political and social action* should be in this orientation, its members clearly share the same values or ideological framework. They were and became known as the "liberals," "do-gooders," and "social reformers," who served as the social conscience of the APA. With respect to any major national or international issue in which the rights, dignity, or well-being of groups of individuals were threatened, the SPSSI Council and its membership would usually discuss whether or not a public stand should be taken, and if so, what it should be, while encouraging the systematic research on the major social dilemmas of the day. It was SPSSI, for example, through the efforts of its distinguished former President, Kenneth Clark, that the behavioral science research literature on the effects of desegregation on black children was organized and used to argue against the "separate but equal" concept in American education before the Supreme Court in the now famous 1954 case of *Brown vs. Ferguson.*

Well over a decade later, the SPSSI Council publicly took issue with the findings and interpretations that Arthur Jensen reported on the intelligence of black and white school children in the *Harvard Educational Review* (1969). According to Jensen s interpretation of his findings, evidence existed of genetic differences in I.Q. between black and white children. A public debate ensued between Jensen and Martin Deutsch, then the President of SPSSI, who advocated the view that inferior I.Q. scores of black children merely reflected their inferior status and existence in a society dominated by its white middle-class members. In this debate Deutsch spoke for himself and not for SPSSI. Yet, concerning the methodological rigor of Jensen's study, what his findings actually revealed, and what interpretations could be placed on them, many psychologists, including SPSSI members and other behavioral scientists, supported Deutsch's position.

When Jensen's study was first published in the *Harvard Educational Review* in 1969, the conditions for the politics of scientific controversy were quickly established. When "Comment" appeared in the *American Psychologist* some three years later, there was little question that some SPSSI members would undertake either informally or formally through its Council to make some response to it. Thus, at the meeting of the SPSSI Council in Honolulu late in August 1972, the Council established a "Commission on the Renewed Assault on Equality." Among other matters, the Commission was to investigate the validity of the charges made in the "Comment," and those implied by the various "beliefs and principles" subscribed to by the fifty signers of the "Comment." Above all the Commission sought to obtain evidence that would lend credence to the various assertions in the letter and resolution that took the unequivocal position that the academic freedom of hereditarians and other biobehavioralists were seriously in jeopardy insofar as writing, doing research, teaching and publishing were concerned. As President of SPSSI. I was asked to be Chairman of the Commission which was comprised of eleven other members seven of whom were SPSSI members and behavioral scientists, and four of whom were biologist-geneticists and not members of SPSSI.

Some characteristics about "Comment" should be clarified. Among the psychologists, biologists, geneticists, psychiatrists, sociologists, biochemists or physiologists who signed "Comment," clearly psychologists dominated the group. The letter and its resolution was submitted by Professor Ellis B. Page, an educational psychologist at the University of Connecticut, and it seemed that the field of psychology was clearly the arena for the controversy explaining the development of human intelligence. Why only fifty people signed "Comment," and whether other distinguished researchers were approached and declined to participate due to ideological or other reasons, was never ascertained. It could be that only fifty signers of "Comment" were sought to obtain the support of a small number of the most distinguished scientists among biobehavioral researchers. Of those who did sign the resolution, the majority are very well known in their respective fields of scientific inquiry. No less than *three* Nobel Laureates signed "Comment." Those who were not as well known as researchers had, however, achieved positions of prominence as journal editors, university administrators, or directors of research centers or institutes.

"Comment" made essentially three points which presumably provided the basis for the five "beliefs and principles" which were the substance of the resolution. First, the hereditarian approach, as expressed in a concern with the influence of biological-genetic determinants on the organization and development of complex human behaviors, has repeatedly been validated by the research of many investigators. Second, hereditarianism is suppressed covertly by preventing those who support it from teaching, doing research, and having their views and research widely published, because it runs counter to the prevailing view of the political militants and those who take an environmentalist position. More generally, it is suppressed because the antihereditarian or environmental view "dominates the liberal academy." Third, the suppression and censure of those who support a hereditarian position take the form of emotional attacks by misquoting and misrepresenting rather than scientific rational opposition. Often these scientists, instead of their evidence, are attacked.

The resolution itself consists of five parts. The first stresses the strength of hereditary factors; the second encourages research into the "biological hereditary bases of behavior;" the third emphasizes the right and scholarly duty of the teacher to discuss hereditary influences in behavior; the fourth deplores the evasion of hereditary reasoning in current textbooks and the failure to give it sufficient emphasis in a variety of behavioral science disciplines; the fifth calls upon all liberal academic groups, e.g. faculties, ACLU, AAUP, journals, boards of trustees, etc. "to insist upon the openness of social science to the well-grounded claims of biobehavioral reasoning, and to protect vigilantly any qualified faculty members who responsibly teach, research, or publish concerning such reasoning."

After a series of meetings in New York City the Commission unanimously agreed to write to all of the fifty signers of "Comment." The full text of the letter begins on page 8. It was dated December 6, 1972, and was mailed some five months after the "Comment" appeared in the *American Psychologist*. The purposes of the letter was first to state the Commission's own positions with respect to the freedom of scientific inquiry, namely, the right of all sides on the issue of hereditary and environment to be heard publicly, but in turn the no less corresponding right of the public to respond to these issues. Second, and perhaps of far greater importance, the Commission asked each signer to provide information and evidence of the various allegations made in "Comment" concerning, for example, the suppression and censure of those with a hereditarian point of view, i.e., who did what and how was it done. The letter sent to the fifty signers had enclosed with it a *copy* of a second letter prepared by the Commission and sent to the *American Psychologist* for publication. This letter announced the formation of the SPSSI Commission, its role in investigating the charges made by the fifty signers, its desire to gather information from the signers. The letter stated that the Commission agreed with the basic position of academic freedom espoused by the signers but explained why the Commission took issue with the way the signers presented their case.

Twenty-four of the fifty signers of "Comment" responded to the letter of December 6 sent by the Commission. The length of their written responses, the degree to which they provided information and documentation, their manifest purposes in writing, the tone of their responses varied considerably. Attempts to elicit more than twenty-four responses were unsuccessful even with follow-up letters after the initial communication. Actually, the Commission received more than twenty-four letters, as some signers who responded initially continued to communicate with me as Chairman in response to my personal letters which attempted to answer questions and criticisms raised by these respondents.

The politics of scientific controversy characterized each of the groups involved in the conflict. Willingly or unwillingly personalities became involved; changes and countercharges are made; biases were expressed; and indeed mistakes were made in presenting each respondent's case. Science is not value free; nor is the stake involved simply one of determining which of two theoretical positions is correct or *the* truth. Truth, particularly in the scientific conflict involved in the attempt to establish it, has many implications extending from the status, influence, and power of particular scientists to the behavior, experience, and indeed treatment of large numbers of individuals in a democratic society. To engage in the scientific process—whether the problem involved is the nature of atomic energy, the causes of the common cold, or the influence of hereditary/environmental factors on human intelligence—is to engage in social communication and social influence. The scientific process involves not only one's scientific peers but also the public at large. There is no ivory tower in scientific research nor was there ever one. Thus, like it or not, all scientists who engage in the scientific process must expect to be challenged, criticized, and even publicly denounced, particularly when *the* truth cannot be ascertained with scientific certainty. However, there are always some scientists who are ready to proclaim they know *the* truth with scientific certainty.

SPSSI and its Commission were part of the political processes involved in the hereditary-environment controversy concerning the development of human intelligence. Many of the fifty signers who answered the first letter from the Commission bitterly complained about—sometimes contemptuously criticized—two aspects about the Commission. First, was its name, the "Commission on the Renewed Assault on Equality,' and second was its composition. The name of the Commission clearly suggested that no matter how impartial the Commission's investigation of the various allegations made might be, its members had apparently already decided the case against the signers of "Comment." Some signers felt the Commission thought they were merely espousing an anti-environmentalist or strongly hereditarian position; why else use the "Renewed Assault on Equality" in the name of a Commission investigating the charges made in "Comment" and in the Commission's accompanying letter? The possible bias of the Commission involved more than the name according to some of the fifty signers. Some of the Commission members, it was pointed out, had been actively and directly involved on many occasions in the conflict with Jensen concerning his research, and SPSSI itself had taken a strong position on the importance of social experience. Therefore neither SPSSI nor its Commission could carry out an "impartial" investigation.

Regardless of the intention of some or all of the Commission's members, these criticisms were not without merit. SPSSI and its Commission were and had been part of the politics of the hereditary-environment controversy. SPSSI, or at least its Council, had taken a very strong position against Jensen's analyses and interpretations when they were reported (1969). The fact that Martin Deutsch and Cynthia Deutsch were Commission members was mentioned many times by various signers of "Comment." Apparently, no conclusions and interpretations, although based on evidence, could be produced by the Commission that would be accepted either by the signers or by even many impartial colleagues in APA who were not nor had ever been part of the controversy.

Therefore, a second letter was sent to the fifty signers. The validity of their criticisms of the supposed objectivity of the SPSSI Commission was fully acknowledged. A direct attempt was made to gain the cooperation of the signers on the basis of changes made immediately in what the Commission's function would be. The full text of this letter begins on page 11. It informed the signers that the Commission had changed its name to "SPSSI Fact-Finding Commission on the Suppression of the Academic and Scientific Freedom of Hereditarian and Biobehavioral Researchers and Teachers." It reaffirmed its intent to collect information directly provided by the signers of the resolution and/or from leads to such information that they might provide. Once obtained, such information would then be given to a group of impartial APA "elder statesmen" who as a Special Committee would summarize, analyze, and interpret the obtained information. Finally, the Special Committee's report would be published by the Commission with its findings made available to all. The Special Committee would give its report to the Commission which, in turn, would make it available to all concerned so that they could make their positions known and thus be heard on an "equal-time" basis.

The Executive Director of the APA, then Acting Editor of the *American Psychologist*, did not immediately publish the initial letter sent to the fifty signers and the SPSSI Commission's letter describing its membership, purposes, and intent. To save space he suggested publishing a single letter giving all of the information provided by the two separate letters including the changes in the name, purposes, and activities of the Commission. This letter appeared under "Comment" in the July 1973 issue of the *American Psychologist*. By this time all the correspondence between the fifty signers and the SPSSI Commission had ceased, and one member, Curtis Williams, a biologist, resigned from the Commission.

WHAT DID THE LETTERS REVEAL?

The intent of the Commission to have a Special Committee of APA "elder statesmen" summarize, interpret, and draw conclusions from the information provided by the fifty signers never came to pass. Perhaps, many factors were responsible: clearly, little new or even relevant information was provided by respondents' letters; no real leads to such information were provided to Commission members for further investigation. Finally, the answers to either the first or second letters by almost half of the fifty resolution signers came from only a few scientists who wanted to cooperate in the fact-finding purpose of the Commission. These few exchanged letters with the Chairman of the Commission and other signers, but they did not provide information. Rather, the letters were attempts to resolve misinterpretations or to defend one's position as a signer or a Commission member. Perhaps what is most telling about all the letters received is that in no single instance did a signer say irrevocably that the respondent (or co-signers) would give facts *and* that approval was given to forward those facts to an independent group to analyze and interpret in the context of the resolution's implications and allegations.

Most of the correspondence is available to be read, interpreted, and evaluted by any behavioral scientist or anyone else. Not all of the letters received from twenty-four of the original signers have been reproduced by SPSSI, because some individuals did not provide the requested release to reproduce their letters for public discussions. Only eighteen individuals signed releases. Although quoted statements in the discussion that follows were taken only from the letters of these eighteen respondents, my analysis and interpretation are based on *all* letters received. Copies of all letters by the eighteen signers and the answers to them by the Chairman of the Commission are all available upon request by writing to SPSSI's national office at the University of Michigan.

My discussion is neither an "official" analysis and interpretation of some independent Special Committee, nor is it the "official" statement of the SPSSI Fact-Finding Commission. When the original purpose of the Commission to collect facts was not achieved, the existence of the Commission ceased, although some members participated in one or both symposia on the hereditary-environment controversy held in New Orleans in the summer of 1974. In its second letter to the fifty signers, the Commission had indicated that symposia should be held to allow all points of view on this controversy to be heard and that signers would be asked to participate. The final action of the Commission, therefore, was to organize these symposia at which Commission members, some of the fifty signers, and other behavioral scientists were asked to participate. This volume, of course, is a result of these symposia.

My analysis and interpretation of the letters is my sole responsibility as the Chairman of the SPSSI Commission who does *not* speak for the Commission. No attempt is made to judge whether or not there is real evidence to support the views and allegations made in the "Comment." A judgment would be an *impossible* task in the first place given the heat of the issue and the different meanings that can be attributed to the same resolution by the fifty people who signed it. Furthermore, just what constitutes the nature of evidence would be equally difficult to decide. Therefore, one of the signers, Robert Cancro, a SPSSI member, explains in his letters what the evidence is and why it may be considered evidence. Unfortunately, neither Robert Cancro nor others were ready to accept what was clearly inferential or circumstantial evidence as direct evidence—at least insofar as the serious allegation of "suppression, censure, punishment and defamation" of hereditarians.

Yet, for all of these difficulties there was and is much to be learned from all the letters received about the validity of the allegations, and about what prompted some co-signers to lend their names to the resolution. Perhaps most importantly, we can learn something about the inevitable politicalization of an unresolved scientific controversy that has very significant implications for the status, well being, and future of large numbers of individuals.

THE LETTERS AND THE EVIDENCE

What kind of evidence was submitted by those who responded to the Commission's two letters? The evidence involves either known accounts or direct experience in which some scholar or researcher (including the signer him or herself) had been verbally abused in a public setting, or threatened in letters and newspapers, harassed through telephone calls, or even in some rare instances attacked physically. Again and again some of the signers described what they themselves had seen happen to Jensen and to others. A small group of signers at the University of Connecticut gave very detailed accounts of what had happened to them and other colleagues, in an atmosphere of fear in which there was, they stated, an inability to do research and speak from a hereditarian point of view because of the possibility of academic reprisals from extreme faculty and student groups. The University of Connecticut's case is a special one. That one of the fifty signers, Robert Cancro, a SPSSI member, signed the resolution with the full conviction that its purpose was clearly and unequivocally in defense of academic freedom, makes a reading of his letters and those of H. Fairfield and David Zeaman enlightening.

It is important to understand fully the response of the Commission to the evidence presented. The events described were real; the evidence was valid. Meetings involving Shockley, Jensen, and others were prevented from being held at given times. Jensen and his family were threatened with bodily harm. Eysenck was threatened, harassed, and later experienced bodily injury. Rosenthal, speaking at a symposium on crime prevention in relation to heredity-environment issues, was harassed, shouted down, and accused of being a racist. Since these events were not in dispute, the Commission as well as SPSSI had long ago condemned such actions as infringements on a researcher's freedom of speech and scientific inquiry. In many of my answers to the signers of "Comment," I indicated my willingness to have signed the "Comment" if the resolution had been confined to threats to the academic, research, and civil rights of those who studied the biological bases of human behavior.

"Comment" raised more issues than that of freedom of scientific inquiry. Reread all the paragraphs of the letter after the first introductory paragraph, and more particularly, the fourth and fifth parts. A critical question to ask about all the attacks against scientists is who made them. Aside from the particular conflicts between Jensen and other behavioral scientists—and attacks were made on both sides—in almost every instance there were extreme radical groups called "Science for the People," "Racial Coalition,' etc. comprised of students, faculty, and unidentified sympathizers. Yet, "Comment" clearly implies that universities, publishers, academicians, and scientists participated in a systematic attempt to censure, punish, suppress, and defame those with an hereditarian orientation. Consider whether the attacks cited—and their validity is not in question—constitute evidence that biobehavioral researchers were being prevented in any systematic fashion from teaching, doing research, publishing, or stating their views within the scientific and the university communities. Parts of the resolution, particularly the fifth, make charges that require a very different kind of evidence than what was presented. The fifth part of the resolution in effect states that groups like faculty senates, professional and learned societies, AAUP, ACLU, boards of trustees, scholarly journals are, by not speaking out, party to the suppression, distortion, and denunciation of the hereditarian point of view by extreme radical groups, and other dissidents. The "Comment" is not deploring a particular group or individual attacking the biobehavioral orientation, but rather—either wittingly or unwittingly—is defining a systematic attempt to suppress, nullify, and indeed eliminate the biobehavioral orientation in the consideration of complex forms of human behavior.

To me, there was not one shred of evidence submitted in the letters and material sent to SPSSI that what the "Comment" stated directly and what some parts of the resolution clearly implied was indeed true. Yet, our analysis has to be correct if for no other reason than the fact that, apart from the harassment and abuse of particular biobehaviorists' research, teaching, and publications involving biobehavioralism were thriving before the publication of Jensen's paper in 1969. They continue to thrive today.

Why then did fifty prominent behavioral scientists sign the resolution? Most of them may have signed the resolution without knowing about the content of the accompanying letter. As for the resolution itself—with or without the letter—like all other petitions involving emotional issues that have become politicized, those who signed may have done so for a variety of reasons. Some respondents admitted they had not considered all parts of the resolution carefully, because the attacks being made were so vicious that as scientists and human beings they had to speak out without quibbling over words and phrases, and without deliberating whether they agreed with everything. A few of the signers were SPSSI members who regarded themselves as liberals, but there were many more who questioned Jensen's findings and his approach. But by signing "Comment," they defended his right to state his findings, carry out research, and take a position on them. (Cancro, Kelly, Rosenthal, Vernon, et al.)

That the viciousness of the attacks against biobehaviorists provoked many to sign "Comment" must have been especially true for Cancro, McLoughlin, Rosenthal and Zeaman who directly experienced such attacks. Cancro—and undoubtedly others– did not see the fourth and fifth parts of the resolution. He signed, therefore, nothing more than what he believed, namely, that he, Jensen and other biobehaviorists were being attacked, maligned, and harassed. The first three resolutions were simply a rallying cry to join others in opposing these attacks and defending his and other scientist's right to freedom of scientific inquiry. Those who were attacking his freedom were in his view "plotting" to suppress and defame him and others like him. The repeated attacks led Cancro and others to an explanation of them based on "plotting," "systematic attack," or "censure."

Cancro's letters and the events at the University of Connecticut and some other universities help to explain why some scientists who signed the "Comment" saw a need for the fourth and fifth parts of the resolution. Much more seemed involved than just isolated faculty members or radical groups attacking biobehaviorists. At some universities student-militant political group uprisings had occurred in the late sixties. At a campus where militant political groups of students and faculty were still highly active concerning equality for all minorities, hereditarians were particularly vulnerable to attack when they discussed minorities and race in terms of heredity. A question arose on these campuses as to whether a faculty senate and an administration could—and indeed should not—step in to prevent militant groups from speaking, influencing, and generally taking a stand. How does one draw the line between censorship and the rights of scientists *and* militants? Clearly, personal threats, physical abuses, and interference with public meetings must and should be stopped, but how is intimidation stopped without causing an intensification of conflict? Even if we agree that abuses can or should be stopped, how are they stopped especially when they involve small "hit and run" extreme political groups?

Is it conceivable then that under these circumstances, some signers of "Comment" saw themselves "suppressed," "defamed," "censured," and punished not just by extreme militant groups but by their colleagues and students and by faculty senates and administrators who were unable to change the situation? At my own university, City University of New York (CUNY), the President of a college, supported by many faculty, students, and administrators, refused to cancel a scheduled debate *he* had arranged between Shockley and others. Yet, the emotion, agitation, and disruption in the auditorium was so great that the debate never really occurred. The President and his administration were severely criticized by some for allowing the debate, and by others for failing to see that it did not occur. In a context of a continuing political agitation, all who are subjected to harassment on a college campus may want to retaliate—or ask others to retaliate—in ways that are no less of questionable democratic and intellectual merit than the ways they attribute to their extreme militant detractors.

Perhaps a third reason exists as to why some scientists signed and supported the resolution. During the years of the Johnson administration, campus unrest, and the Civil Rights Movement, the environmentalist view may have been given greater attention than the hereditarian approach in textbooks, courses, and in research. (I am not entirely convinced that this is true, because many papers and books were published and many public presentations were given by many of the signers before and during this period.) To what do we attribute the focus on environmental views? Was the inattention then to the hereditarian approach, as is implied in the fourth part of the resolution, a systematic attempt to suppress, censure, and defame this approach? Or is it, as seems more likely, the swinging of a pendulum of theoretical and empirical preference in which a particular faculty in giving courses and doing research, reflect particular orientations, and receive attention in greater numbers? Are the number, importance, and indeed power of other teachers and researchers consequently lessened?

To label this suppression and censure in the manner of a witch hunt is once again a judgment that gives evidence of "causal contamination" by the politics of scientific controversy. The abuses, the agitation, and indeed the infringements on the academic freedoms of hereditarians by extremist political groups—with some environmentalists perceived as the architects—are now seen as parts of a master plan to exclude hereditary reasoning from research, teaching, publications, and faculty representation in the university and other academic settings.

Not only is this "master plan" in which a given set of effects are associated with irrelevant if not erroneous causes, but the history of behavioral science disciplines—particularly psychology—in the United States is being ignored. From 1930 to 1960, academic departments, curricula, books, and journals usually reflected and indeed boasted a given theoretical orientation, clearly dismissing other orientations as unworthy of attention. Were there not departments in which Hullian learning theory reigned supreme? Didn't some schools have departments in which experimental psychophysics held such dominance that social, personality, and developmental psychologists either had no place or if they did, they barely achieved second class status? Many psychologists today can "regale" the various signers of "Comment" with stories of how tenure was won and lost depending on whether one's approach, articles, and books revealed—other things being equal—the right kind of dogma and no signs of heresy. Hereditarians were as subject to such restrictions as were gestaltists, Skinnerians, or Piagetists. While many of us have condemned this kind of "theoretical authoritarianism" whether it was Hullian learning theory, Lewinian theory, or orthodox Freudianism, we can hardly be accused of "suppression," "punishment," "defamation," or "censure" that result in our violating academic and scientific freedom. Attacks made on those who do not subscribe to a theory or approach may sound like a systematic attempt to control knowledge, research, and the rewards of behavioral scientists. But it is as difficult to conceive that such a "plot" exists as it is to see the publication of books using the Skinnerian approach to education, therapy, and social problems as a "plot." That people happened to be hurt while giving emphasis to a particular theoretical or methodological emphasis in a given department was indeed the case. Nor should it be denied that in some instances scientific prejudice was deliberate whereas in others it resulted unwittingly from the frailty of seeking theoretical congeniality in teaching and research among professional peers. In terms of the essential purpose of the *academy* of the university, witting or unwitting scientific prejudice must be condemned. However, in no sense can it be equated with the systematic disenfranchisement of women, blacks, and other minority group members that continues to pervade scientific life in America.

CONCLUSIONS

In summarizing the letters and information SPSSI received, there was considerable evidence of attacks on and abuses of particular hereditarians. These attacks are to be deplored, because they represent violations of the academic and personal freedom of behavioral scientists to do research, write, and teach scientifically and intellectually viable concepts. However, there is no evidence, as implied by the resolution and clearly indicated by the accompanying letter, that biobehaviorists have been systematically disenfranchised as "heretics" so that they cannot teach, do research, publish, or achieve recognition for their point of view.

Did the three documents which begin this volume and the SPSSI Commission resolve a scientific controversy? Probably not if one seeks to bring an end to a scientific controversy or expects to triumph by some intellectual breakthrough that convinces many behavioral scientists or biobehaviorists of error. Such outcomes are expected only if one assumes what was at stake was the resolution of a scientific conflict. Such was not the case, nor was it ever the case beginning with the publication of Jensen's study. What is at stake are the *politics* of scientific controversy. Regardless of the intentions of the writers of the resolution or those who signed it, "Comment" was and is a political document. It resorted to an emotional appeal, drew conclusions on an *ad hominem* basis, and in the letter or the resolutions identified as culprits, environmentalists, faculty senates, foundations, and organizations. It was alleged that the culprits were part of the attack or aided and abetted it by not speaking out. Although the attempt by the SPSSI Commission to get the facts was abortive, "Comment," as a political document, had to be answered. There were and are two unresolved matters of critical importance: first is the welfare of people most affected by the original controversy, the millions of individuals we identify as racial minorities, and second is the value of science itself as an approach to knowledge.

As for the first, the case against the resolution is easily made. We have no doubt that most signers of "Comment" are not racists. That it was signed by Jensen and others who publicly support his views and interpretations of the role of heredity in the difference in I Q. between whites and blacks indicates that they are inadvertently lending support to Jensen's views and interpretations. Certainly Jensen's right to do his research and present his views must be vigorously defended, *but not in a resolution that defends this right without clearly stating that his views are not being endorsed.* It does little good for some signers to write to SPSSI that they do not subscribe to Jensen's kind of research, his findings, or his conclusions. How does one communicate this to the many thousands of behavioral scientists and non-scientists who read "Comment?" Jensen's name listed with three identified Nobel laureates could be interpreted as meaning that indeed racial differences in intelligences are primarily rooted in hereditary-genetic factors. In more than a few letters SPSSI received, the need to defend Jensen's right to research, publish and speak, became a defense of the importance of hereditary influences as revealed by his findings.

It is no less important to point to other aspects of the political character of the controversy revealed by "Comment" and in some of the replies received by the Commission. *No where* in the resolution are the rights of *all* scientists, including environmentalists, to do research, study, and teach espoused. What happens now to environmentalists? With an apparent national inclination to the political right, it may well be that they will replace the hereditarians as the "injured party." On some campuses, environmentalists like Martin Deutsch and others have experienced verbal abuse, threats, and other indignities. One signer of "Comment" in a telephone conversation, revealing just how political the conflict is, said, "If environmentalists have also been attacked, I just haven't heard about it, and until I do there is nothing I can say about it." Perhaps not surprisingly, in some letters and telephone calls I received, the SPSSI membership was denounced as SDSers and "fellow travelers."

As we have already suggested, what is at stake is the well being and the futures of many Black Americans if not all Black people. Scientists, particularly behavioral scientists, have to learn that the efficacy of their approach to knowledge will depend on the extent to which they recognize their social responsibility as scientists—whether they are hereditarians or environmentalists. In a highly technological, complex world in which instant and easy communication has become a way of life, what a scientist says and does as a scientist may have profound effects on social and public policy and consequently on the well being of many human beings. I summarized my own thoughts and feelings about scientific responsibility in a letter to Francis H.C. Crick, one of the three Nobel Prize laureates who signed "Comment:"'

> I repeat what I said in my letter. Scientific freedom is not just a right but a *responsibility*. This means that the meaning and use of research findings is an important aspect of the responsibility....Thus, the *manner* of presenting findings, particularly when they can have immediate consequences for human events is critical. I feel exactly the same way about scientists who even in writing for the scientific community tend to forget about the *manner* they use in stating what they found.
>
> As for your view that further well-designed experiments could decide the matter on the inheritance of I.Q. . . . . . The AAAS and the National Academy of Sciences and other groups have made it clear the experimental answers to questions are not possible on either side. That's the point, neither side can ever make an empirical case as far as matters stand *now*. However, this does not mean that geneticists should not be allowed to do such research, receive money to do it, or in any way be prevented from following whatever position they have. I would hope, however, they would—as well as the environmentalist—act as *responsible* scientists when it comes to what they say and do.
>
> To be personal, I must say I was shocked by the next to the last paragraph of your letter . . . your implication—that there is suggestive evidence on the inheritance of I.Q. and therefore we had better not force these people into aspiring to what they cannot achieve—absolutely appalls me. Would you as a first-rate scientist really be willing to take that stand on "suggestive evidence?"'Isn't it possible that it would be better that they strive in the hope that they may achieve, than that they be told they cannot achieve because they are doomed by heredity to a lower status in our society? I don't want to take either position. I just want scientists to realize that there are many questions they have not answered, and therefore, not to act, talk, and influence the lives of other people, as if they have.

REFERENCES

Aronson, L.R., E. Tobach, D.S. Lehrman and J.S. Rosenblatt (eds.) (1971). *Selected Writings of T.C. Schneirla*. San Francisco: W.H. Freeman Press.

Cohen, David (1972). "Does I.Q. Matter," *Commentary*, Vol. 53 (4): 51-59.

Deutsch, Martin (1969). "Happenings on the Way Back to the Forum: Social Science, I.Q. and Race Differences Revisited," *Harvard Educational Review*, Vol. 39 (3): 523-557.

Deutsch, Karl and Edsall, Thomas (1972). "The Meritocracy Scare," *Society*, Vol. 9 (10): 71-79.

Eysenck, Hans (1971). *I.Q. Argument: Race, Intelligence and Education*, New York: The Library Press.

Herrnstein, Richard (1971). "I.Q." *The Atlantic*, Vol. 228 (3): 43-58.

Jensen, Arthur (1969). "How Much Can We Boost I.Q. and Scholastic Achievement," *Harvard Educational Review*, Vol 39 (1): 1 23.

Tobach, E. (1972). The Meaning of the Cryptanthroparion. In *Genetics, Environment and Behavior*. L. Ehrman, G. Omenn and E. Caspari (eds.) New York: Academic Press, pp. 219-239.

THE SCIENTIST AND HIS FINDINGS: SOME PROBLEMS IN SCIENTFIC RESPONSIBILITY

Abraham Edel

Current controversy over race in its relation to intelligence runs the gamut from questioning whether there should be any research in the area to proposing specific legislation on the assumption that definite results have already been achieved. For example, Bodmer and Cavalli-Sforza, at one extreme, conclude that "for the present at least, no good case can be made for such studies on either scientific or practical grounds."(1) Shockley, at the other extreme, not merely concludes with confidence that "perhaps nature has color-coded groups of individuals so that we can pragmatically make statistically reliable and profitable predictions of their adaptability to intellectually rewarding and effective lives,"(2) but goes on to propose a program of offering bonuses for voluntary sterilization.

In the light of this controversy, we shall have to discuss—if we wish to pinpoint even in outline the range of the scientist's possible activity: 1. choice of research topic, 2. modes of doing research, 3. modes of presenting results among scientists, whether of experiment or survey or theory, 4. mode of publicizing results and ways of engaging in controversy about them, and 5. the scientist's participation in the formation of policy, whether social or governmental (down to specific legislation).

Since the controversy is today occurring against a background of assumed common values about freedom of inquiry, any account of responsibilities must first make sure we understand this libertarian framework. Briefly, it centers—in the context of scientific work—on two values: one is truth and the other is freedom of thought, inquiry and expression. Some see liberty as an absolute right, others regard it as the appropriate structure for a society determined to achieve the material and cultural progress essential to a growing well-being of its people. The differences may not be relevant to our inquiry, chiefly because both tend to assume that freedom of inquiry is the best means for ensuring the discovery of truth. Concerning truth itself there are also two views: some regard its pursuit as an absolute value; others see it as a very high value to be pursued but not at any price.(3)

A simple appeal to libertarian rights may not itself be enough to locate the points of controversy. We have to recognize that the theory of liberty in our tradition is not simply the proclamation of a single principle of freedom but has involved an often precarious balancing of three elements: 1. a maximum of individual liberty, 2. a framework of rational discussion in resolving disagreement, and 3. a specification of limits directed to self- and social-protection.(4)

Of these, the first obviously needs no discussion. The second element is perhaps too strongly formulated, for it may suggest rationality in the discussion itself rather than the necessary conditions for rational discussion. The necessary condition for rationality is the willingness to listen and to be heard in turn; it is a framework of civility which is something more than keeping quiet and something less than rationality itself. The third element is the most difficult in our tradition, because limits proposed often tend to become seats of dogmatic authority through which the established order holds back change. Generally, freedom of thought and expression does not guarantee complete freedom of action. Concepts of sedition, license, clear and present danger, pornography, etc., have at various times been offered in different fields to pinpoint borderlines of action, to suggest limits, and to legitimize paternalism. Traditional experience in these concepts has not been very happy. "Clear and present danger" was voracious enough in the Smith Act cases to trespass on the first amendment, and it has taken a long while for us to recover. We have never gone so far as Harold Laski once suggested (criticizing the decision in World War I cases)—that dissenters should even be allowed to hand to soldiers actually embarking for the war front pamphlets which urged them to refuse to serve. At least the British soldier, he said, was so indoctrinated that if the pamphlet could persuade him, its arguments were probably correct! In any case, whatever the complexities in setting limits, the rights of self-protection and social protection are clearly as firmly established as the rights of freedom of expression; both have their place in basic rights. The real issue arises when they are in conflict; one must then decide whether dangers are real or people are conjuring up dragons.

We turn now to the question of the scientist's responsibility with respect to each of the five topics in the light of the libertarian framework. In some cases the treatment can be brief, in some fairly definitive, in others at best suggestive of the complexity of the problems.

CHOICE OF RESEARCH TOPIC

As a *prima facie* matter, there seems little reason for not allowing full scope to the liberty in a scientist's pursuit of any intellectual interests that he may have. Society has enough positive incentives to attract a sufficient number of scientists to what may be felt as socially urgent. Moreover, the scientific maverick has on many occasions been a source of fruitful intellectual mutations, even beyond keeping orthodoxies on the alert. Still, it is conceivable that specific scientific responsibilities to avoid dangers of great material and social harm may give rise to an obligation not to enter on specific researches at a given time. The first temptation in the liberal tradition is to rest on the value of truth and say that responsibilities will emerge only in the later steps of publicizing and applying one's findings. Whether it is possible to control these subsequent steps is itself an empirical-scientific question concerning the character of communications and the state of the existing society. Therefore, the general responsibility of the scientist to become conscious of and to consider the consequences of research in an area can be recognized at this point. The types of considerations that should enter into his judgment may be gathered later, but in the libertarian tradition the decision is the individual scientist's own.

A second general responsibility in choice of a research topic is fairly obvious. Although a scientist may generally not be concerned with why he finds a research problem interesting or attractive, he should try to determine, insofar as possible, the extent to which his motivation may involve him in possible ideological conflict. This determination is clearly important in an issue like race and intelligence, in which both racist and utopian attitudes may be found. There is no reason why a scientist should not investigate a field because he hopes the results will be of such-and-such a sort. (We may pass independent judgment on his motives.) However, he should where possible be aware of his hopes so that he will thoroughly test his findings.

MODES OF DOING RESEARCH

It does not follow immediately that any scientist who has thought through any interest and projected research should be absolutely free to pursue it. Any control over research, whether society's or one's own, is guided by the major countervailing factor in the libertarian framework of the need for protection. For example, a particular biological project might require producing a culture that could, without control, endanger many lives; decisions in research would accordingly depend on safeguards available. The research might involve experimenting on human subjects; the moral problems here have recently become prominent in medicine, psychiatry, and even experimental psychology.(5)

To recognize that freedom is not absolute is not enough. We need a complex ethical inquiry into responsibilities, types of legitimate checks, location of burden of proof, areas of legal coercion or administrative obstacles, and dangers of inhibiting discovery through imposing present scientific orthodoxies dogmatically. Whether there are general answers or only methods for approaching particular situations is not itself clear. Thus we seem to have reached an agreement on ruling out experimentation for biological warfare, and it seems unnecessary to discuss the mad scientist in fiction who wants to perfect a cobalt bomb to end life on this planet. Norbert Wiener was only one outstanding example of a scientist who turned away from research that would contribute to war, but his was a personal moral decision. The social moral problems here are what kinds of research should and should not be restrained, i.e., research to be legislated against, research to be forbidden at public expense, research to be discouraged, and, of course, the kinds of research to be positively stimulated.

Several points about countervailing considerations to absolute freedom of research are worth noting. First, these considerations refer to the impact of the research itself, not to the applications of results, which raise subsequent questions—for example, early fears that computer research, when applied, would produce unemployment. Second, the considerations involve both the present state of knowledge and present patterns of value. With respect to our knowledge, we cannot assume infallibility. With respect to values, coercive restriction on investigations can be considered only when serious social dangers to life and health may be involved. We cannot impose on research a special morality or the values of special class or religion. Third, the extent of the need for controls reflects the state of development of science itself and its role in society. There is a difference, for example, between the times when science was the work of isolated individuals whose investigations seldom affected the character of social life, and contemporary times when science, including basic research, is a major dynamic factor in the functioning of our societies and may generally alter the character and quality of life. Fourth, because of the broad nature and role of scientific research today, indirect controls will come from the need to allocate scarce resources. Finally, there is the moral responsibility of those who decide on controls to avoid dogmatic imposition of special orthodoxies of knowledge and value.

PRESENTATION OF RESEARCH RESULTS

It might at first seem that the presentation of results among scientists (not publishing to the general public) could be taken as a matter of course: one completes research and lets other scientists know about it. (I am not referring here to problems of top-secret classification which inhibits scientific communication in some areas; that is serious enough, but not relevant here.) The belief that one can simply tell, even in technical language, what one has discovered, is misleadingly oversimplified. There are many different kinds of traps to be avoided. Some are external questions, such as the timing of publication. Attention has recently been directed to a tendency to rush into publication, because, in the current contraction of grant funds for research, a competitive advantage accrues to those who in effect advertize through publication promising directions of inquiry. Such publication is unethical when the results are not secure enough as yet to support the premise.(6) More serious are the internal questions of the adequacy of data and the relation of interpretation to data. In the psychological and social sciences, the phenomenon of conflicting schools of thought and theory often leads scientists to interpret research without regard for possible alternatives, and consequently without stating how tentative the enunciated results may really be. It is, therefore, a basic responsibility of the scientist in reporting his results to distinguish between data and interpretation, to exercise the greatest methodological care about data and about modes of manipulation.(7) I am not referring to the difference between a good and a bad scientist, but to the degree of sophistication among good scientists themselves.

The extended controversy on the question of race and intelligence since 1969 when Jensen published his article in *Harvard Educational Review*(8) has demonstrated the tremendous diversity of issues—logical, conceptual, evidential, cultural, politico-social—that are involved in the assessment of his findings. For example: 1. Logical issues, such as whether one can go from individual heredity as a source of individual differences in intelligences to heredity as a source of group differences. Heritability applies to populations, not traits or characteristics alone. 2. Conceptual issues such as how far the interaction of genetic and environmental variables at all stages in a person's development makes the mathematical model drawn from simpler traits in the study of population genetics inapplicable in the case of intelligence. In general, the feasibility and utility of distinguishing between heredity and environment. 3. Methodological issues about the meaning of intelligence and the validity of its measures, the controversial aspects of the belief in a constant general factor, the debates about what the tests test, about the way in which tests are constructed and items revised, and so on.(9) Similar issues about the concept of race itself and the bases of race classification.(10) 4. Evidential issues about the reliability of focal data—e.g., the controversy about the study of twins. (Certainly something has gone scientifically astray when the same material is treated on the one hand as a firm factual basis for definite conclusions about heredity in the I.Q. and on the other as involving sloppy neglect of interfering factors.)(11) 5. Cultural issues about the role of cultural and sub-cultural factors in intelligence tests and testing in cross-racial studies.(12) 6. Sociological issues about the extent to which the tests reflect middle class selection of capacities to fit into and continue the patterned needs and values of the Establishment.(13) 7. Politico-social issues such as the history of ideological use of theories of racial superiority and inferiority in economic exploitation, colonial struggles, immigration policies, etc. 8. Issues of synthesizing to get a general conclusion out of a diversity of evidence of differing weight—how far one can establish the general strength or tentativeness of the theoretical position as a whole, including the problem of weights and emphases.(14)

A proper responsibility in this controversy is of course relative to the broad state of knowledge of the issues involved. Suppose a scientist surveying arguments on race and intelligence completely ignored the anthropological work on the ethnocentric elements in testing. He would definitely be failing in his scientific responsibility, because this work in anthropology is long-standing and could be assumed to constitute a standard caution. On the other hand, he might ignore currently emerging sociological critiques, which, if sound, might show where he erred, but might not necessarily show that he was irresponsible as a scientist. Suppose he failed to consider published experimental studies of the dependence of I.Q. results on the race of the people administering the tests or the mood of people taking the tests. His failure might be culpable neglect of evidence, if he treated the "fixity" of the I.Q. as an established item of knowledge no longer to be questioned. Suppose he attributed an ideological motivation to opposing arguments in order to avoid confronting them. He could be considered irresponsible unless there were other (e.g., empirical) sound grounds for thrusting the arguments aside. Finally, he might level his charges of ideology against those disagreeing with him and ignore possible ideological commitments on "his side." Such an argument would be naive.

Clearly, there are various degrees and shades of responsibility and irresponsibility in the declaration of scientific findings. Their determination is a matter of concern for the specialists in the field and for those in neighboring fields upon which the findings may draw or in which presuppositions of the research may lie. As a non-specialist, I cannot presume to judge whether in the Jensen controversy there has been a high degree of scientific irresponsibility.(15) However, it does seem to me that areas of science, especially in human affairs, can no longer coast on an attitude of "I say what I believe because I believe it." If for no other reasons, the institutional uses of psychology and social science will press responsibility on them; for example, findings in psychiatry will be tempered by studies in law, and research in psychology will be re-evaluated by research in education.

PUBLICIZING RESULTS

Clearly the scientist has some responsibility for the way his ideas and research results are publicized. At the very least, he ought to correct popular misinterpretations and misuses of his research, even if he himself does not write for the mass media. Especially today, when scientists have a widespread influence on the public, it becomes important to consider what responsibilities they have for the fate of their findings in public or semi-public media.(16) Whether he or others are writing, the research scientist should make sure that opinions are qualified by degrees of evidence, alternative conceptions are noted as well as the extent to which results are being applied from a domain the scientist has explored to one which he has not. For example, doubts may be raised whether responsibilities are being met in many popular articles by ethologists on aggressivity, by sexologists on the proper relations of the sexes or on the nature of women and the women's liberation movement, and by psychologists on parent-child relations.

In December 1973—to take a complex illustration—the American Psychiatric Association altered its position of nearly a century and declared that homosexuality is not a mental disease.(17) The import of the change, which technically simply removed homosexuality from a list of mental diseases and appeared to be a revision in nomenclature, was not wholly clear. What the public learned in newspaper reports, television programs, interviews, and discussions among psychiatrists was something like the following: 1. The change is not just yielding to the pressure of homosexual organizations. 2. Because many homosexuals are satisfied with their conditions, there is no point in branding them as mentally ill or forcing them to think of themselves as freaks. 3. There are not enough specialists in homosexual treatment to cover the whole field. 4. Only those homosexuals whose sexual orientation distresses them are suffering from a disturbance.

As far as wanting to know whether the APA's previous position had been incorrect and was now being corrected, the public knew as little as when a press secretary announced from the White House that a previous statement was now "inoperative." The American Psychiatric Association at the same time urged the removal of all legislation penalizing sexual acts performed by consenting adults in private and full civil rights and protections for homosexual citizens.

What would a responsible relationship between scientists and the public have required on the theoretical state of the question raised by the APA's revised position? The Association might have issued something like the following statements: 1. There has long been controversy on the basic theoretical issues underlying homosexuality, both on the biological and psychological aspects. One view treats human beings as having an initial general capacity for sexual response, which becomes varyingly directed in individual development; the other as having an initial capacity for specific heterosexual mating, which becomes diverted (or distorted or crippled) in development and seeks substitute expression. On the psychological side, the first view regards the absence of heterosexual interest like any unactualized capacity, the second as a continuing drive incapacitated by psychologically generated fears. The scientific evidence between these hypotheses is not decisive, so that different analogies can be used and examples found for which each appears to hold. 2. The concepts of health and disease, of normality and abnormality in these domains have large components of value judgment, so that some scientists have even urged that the previous classification was nothing more than an imposition of cultural values through a dubious conception of normal health. Although these extreme formulations are unacceptable to most psychiatrists, the possibility that part of the traditional theory of homosexuality in psychiatry has these value components is a proper hypothesis for evaluation. In addition, there have been historical changes in the medical model, which would have different effects on the concept of mental health. 3. Psychiatry has been unfortunately beset by different schools, each of which has asserted dogmatic truth rather than thought of itself as finding part of the truth. 4. The social background of repression in sexuality has been weakened by social movements of liberation. It is therefore appropriate for psychiatry to be wary of diagnostic branding which impedes an individual's decision about his way of life.

Would the public understand these statements and would an ultimate resolution be seen as a consequence of theoretical and practical trends, with a recognition of the inherent probablilism in most scientific theory and the relative aspects of nomenclature and classification? The public would at least be treated with respect, although its automatic subservience to the psychiatric establishment might be weakened.

In a similar way, any responsible statement publicizing the recent controversy on the relation of race and intelligence must somehow make clear all the kinds of issues involved, together with the paucity of definite evidence, the tentativeness of the conclusions, and the present disagreement among experts. It should separate and define issues rather than treat the whole question as a single package. It might well enter into the history of the problem, showing on the one hand its ideological entanglements and on the other how scientific developments like those in population genetics opened new lines for consideration. In addition to indicating the difficulties as well as the hopes in the investigation, any statement should emphasize the importance of the presuppositions concerning the meaning of "intelligence" and "race" that underlie formulations in the controversy. Most important, perhaps, from the point of view of the history of science, a statement should consider the possibility that formulations will themselves undergo change as science advances and should raise doubts as to whether present formulations may not be an inadequate basis for present and future investigations. Finally, there is no public place for the dogmatism that jumps to a reconstruction of policy in practical fields as if it were a deductive conclusion from an established thesis.

If such cautions are observed, public controversy about a scientific thesis has some chance of maintaining a scientific character. Such cautions were not observed in the public controversy over intelligence and race; it soon became embroiled in charges of ideology and even of conspiracy to repress scientific findings. The charges culminated in a formal statement, "Comment," signed by fifty scientists and published in the *American Psychologist* of July 1972. The rapid development among professionals of this controversy, carried into a public forum, makes essential the analysis of scientific responsibilities in controversy. As to the controversy itself, it has ceased being a scientific exercise and has become a social battle fought with social instruments.

It is worth briefly examining the statement by the fifty scientists on behalf of the study of hereditary influences in human abilities. How far as a specimen of public controversy did it promote the scientific goal of getting strictly scientific solutions to scientific problems as well as scientific understanding of ideological problems? I has, first, a strangely mixed sense of whom it is addressing. If it addresses academics, then why does it appeal to the American Civil Liberties Union? If it is for the general public, then why is there no clarification of the central, crucial distinction between the study of heredity in individual differences and that in groups or races? As an indication of the lack of clarity of the statement, one signer later wrote that the resolution dealt solely with behavior and heredity and took no position with respect to race differences.(18) Further, the statement is prefaced with a background account including the persecution of scientists from Galileo to Mendelian biologists. There is reference to the attack on Einstein in Nazi Germany, yet no mention of Nazi Germany's use of biology for racial persecution and genocide. This is a glaring omission in a statement which supposedly attacks ideological blocking of science. It is sad that this statement by eminent scientists against ideological interference in science was interpreted in its turn as another ideological weapon, but it is scarcely surprising.

Two general shortcomings in the statement prevent it from helping to clarify the confused intellectual-social situation of the controversy and are responsible for its tone of outraged innocence. One concerns the history of science, the other, the character of academic life today.

First, the statement does not distinguish sufficiently between an external attack on science and an internal scientific controversy. It says in effect that in history science has always had to face external intolerance from Galileo's time to our own, and consequently men of good will should rally to the defense whenever science is attacked. This overlooks the marked difference between the time when science was an undeveloped enterprise itself, and the present day when science is powerful and influential. External attacks have not, of course, come to an end, for new sciences arise and often tread on different social interests. However, the model for ethical analysis in Galileo's trial is no longer the major relevant model for analyzing internal controversy like the role of heredity in intelligence. To develop an appropriate model for the latter is precisely equivalent to delineating the responsibilities of scientists in controversy and in social action.

The second shortcoming in the statement is an insufficient refinement in understanding academic orthodoxies. It is important to distinguish between social motivation intruding into the judgment of scientific beliefs and orthodoxies in philosophical or scientific beliefs themselves. The different and often conflicting schools of thought legitimately represent different conceptual frameworks that can be assessed by their relative success or failure in advancing a field. They become less legitimate when they are enshrined as orthodoxies in academic departments and dominate appointments and publication in controlled journals. Yet such domination does not add up to conspiracy to suppress opposing views; it is usually too overt to be conspiratorial and expresses itself in judgments of worth and lack of worth! Such a process has often been found in the conflicts of behaviorism and psychoanalysis in American psychology, of analytic movements and phenomenology in Anglo-American philosophy, of mathematical and institutional economics. If "hereditarians" feel shut out from academic expression, the door is surely less firmly barred for them than for psychoanalytic theorists. Imagine publication of a statement by fifty psychoanalysts about some behaviorist psychologists as persecutors, or a statement by fifty Marxian philosophers on academic discrimination and misrepresentation of their perspective.(19)

A scientific approach to different schools of thought is largely a task of understanding the differences and analyzing their diverse components in order to make comparison profitable and the interplay of differences useful. Instead of breaking off communication there should be analytic dialogue. Scientists should assume responsibility, in the public presentation of scientific findings and in controversy about them, for keeping the spotlight on the scientific character of scientific decision and, when ideology is involved, for turning a scientific light on ideology itself. Scientists should understand what is involved in ideological struggles and not be overwhelmed by them. A scientific statement should not be shrill against the shrill or a hoot against the hooters. The scientist has an obligation to understand why the hooting occurs, not to take the stance of injured innocence.

It is, of course, generally agreed by scientists and intellectuals that, whatever side they have taken on a controversy, the framework of civility is to be maintained; viewpoints are to be reasonably discussed, not shouted at. SPSSI and its officers are just as determined as the fifty scientists in their condemnation of those who disrupt meetings or interfere with the orderly presentation of argument.

Such a general attitude rests on at least three bases. First, libertarians recognize the central role of civility. Second, they fear that any violent interference with freedom of expression will grow from shouts to demands for economic reprisal to threats of physical assault, to terrorism. Third, they generally assume that non-civility springs only from irrationalism and therefore undermines the foundation of a scientific outlook.

What reason leads some people today not merely to engage in conduct disruptive of the framework of civility, but to claim that the disruption is rational? Their claims must be understood, not merely ignored or dismissed. Indeed, the advocates of disruption would claim not to be abandoning the libertarian outlook but to be invoking its component of self and social protection. They also assume that the application of coercion can be controlled and will not result in terror. For our present discussion it is not relevant whether the coercion be private action or economic sanction or governmental penalty following on legislative prohibition; which is to be used would have to be argued separately.

War is the only clear example in our tradition when civility is unhesitatingly abandoned. We do not give freedom of expression to an enemy; we jam his broadcasts without examining the truth of what is being said. We use deliberate and controlled violence to frustrate expression. In peace time attempts to impose a legal limit on free expression have been directed, for example, against communists, as in the McCarthy period. The theoretical argument for this kind of coercion has been that those who will not allow liberties when they are in power cannot expect them when they are not. (The same argument was used against Catholicism in the 17th century.) Generally our tradition has rejected such limitations by trying to link repressive measures to action rather than expression. An interesting unsuccessful attempt to legislate against expressions of belief was a bill introduced shortly after World War II to ban from the mails materials that would incite to racial hatred. The argument over the bill split the then broad alliance of liberals and the politically left. Although some labor organizations endorsed the bill, the American Civil Liberties Union opposed it. The secondary arguments against the bill were persuasive: the legislation would give too much power to postal authorities; it could be misapplied (for example, anthropology text books containing analyses of racial equality could be banned in those states where segregationists were prone to violence); if Congress was unready to support bills for removing discrimination in employment at that time, it should not be trusted to legislate on matters of freedom of thought. The most moving argument for the bill came from veterans who had recently seen Hitler's concentration camps and gas chambers. They argued that the sacrifices of the war had been made to overcome racial violence, and it was folly to allow comparable views to be promoted through the United States mail. To the veterans, the bill was not a question of the freedom of speech or inquiry; it was a question of fighting a war.

Recently, private action to limit expression was seen in the revolt on the campuses against government spokesmen defending the Vietnam war. The action was effective, for it became almost impossible for a governmental representative to appear on a college campus with any assurance that he could present his views. Those who engaged in or encouraged limiting discussion argued that debate within the framework of civility would be ineffective against a pattern of deliberate lies, undisclosed invasions, and designed deceits on the part of the government in power. They believed that their conduct shortened the war. It helped change the climate of opinion, influenced Lyndon Johnson's decision not to run again for the presidency, and helped force Richard Nixon to slow down and eventually to end the war. They saw their moral choice as between abandoning civility to save Vietnamese and American lives and continuing civility at the cost of continued immoral war. The crux of an evaluation of their position would thus seem to be not whether civility was morally violated, but whether their analysis of the situation as a whole was correct: had the policy of the administration really brought the country to the point of war on the home front?

Arguments to justify the abandonment of civility in the controversy on race and intelligence seem similar to those used in the Vietnam war debate. Their advocates recall the depths of frustration that almost destroyed American cities a few years ago and the continuing battles for integration in jobs, housing, and education. They point to the fate of programs of attempted assistance to the disadvantaged and might even refer to *Life* (1970), quoting Daniel Moynihan, who as a White House advisor, said that "the winds of Jensen were gusting through the capitol at gale force."[20] They would trace the ideological role of beliefs about racial inferiority in repressing blacks and other ethnic groups in American history, trace its penetration into intellectual disputes, pinpoint the racial blind-spots in the history of I.Q. testing, give a social explanation of the precise timing of the renewed investigations in an atmosphere of contracting economic opportunity, and show how intellectual complexities were overriden to produce dogmatic conclusions tied immediately to reactionary social policies. They see the central issue not as freedom of speech or inquiry but as preventing pseudo-scientific ammunition from being circulated to the racist enemy. The supposed intellectual discussion is to be regarded as no different from enemy broadcasts during wartime. Although they are breaking with civility, they believe they are acting rationally rather than attacking reason or science.

How can scientists, for whom a framework of civility is scientifically important, deal with such a position? They can at least assume a greater collective responsibility to stop their scientific findings from being used as ready weapons of war on the social-intellectual scene. We have seen what this entails in ways of publicizing scientific views and findings and in analyzing controversial questions. It also entails a deliberate scrutiny of projects and findings that are likely to become involved in social conflicts. Scientists can presumably distinguish, better than the lay public, those projects which are manifestly pseudo-science, those that are speculative and not open to determination in the foreseeable future, and those that are scientifically feasible. Scientific agreement is not perhaps always to be expected, but if a debate about a project or finding is responsibly publicized, there may be the same result as if scientists always were given "equal time" whenever ideological uses of their findings got on the media—particularly if the debate is accompanied with a scientific analysis of the ideological uses themselves. Such efforts will not, of course, defuse social conflicts themselves, but they may help defuse the use of the war model in relation to scientific research.

PARTICIPATION IN THE FORMATION OF POLICY

The responsibilities of scientists in relation to policy and social action and the impact of all stages of their work on the social scene should not be considered as an exceptional or crisis matter. It is already time that scientists face these responsibilities as a continuing collective task.

A number of arguments are, however, sometimes offered as to why scientists should not be concerned with the impact of their findings on social policy. These include: the value neutrality of science, the specialization of the individual scientist, the ideal of progress with the expectation of compensating consequences, and the unpredictability of consequences including the prevalence of unintended consequences.

The argument for the value neutrality of science no longer has the strength it did when science was a small and uninfluential endeavor. Now scientific stands do play a consequential part in social policy. The psychiatric reversal on homosexuality described above was not a merely theoretical scientific matter. The previous classification of homosexuality as an illness had been related to laws and practices which resulted in jailings, ostracism, and possibilities of blackmail. The revised classification was an influence for liberation. As a minimum, the scientist has some responsibility for knowing whether his findings are likely to have a social effect.

The scientist sometimes claims that specialization excuses him from social responsibility. "I am a psychologist dealing with heredity and intelligence," he might say, "I am not a sociologist to reckon on whose passions will be influenced or an educator to balance values of the society in educational programs."(22) If so, let this scientist forego making policy recommendations as if they followed purely from his findings alone, or let him enter into an interdisciplinary project with sociologists and educators. Most important let him remember the dangers of others misusing his results, try to foresee where misuse may happen, and do his best to guard against it.

The ideal of progress is sometimes taken to assure the scientist that any further advances in knowledge will be for the good of mankind, and if consequent social dislocations do occur, people will find compensating modes of action. Therefore, scientists did not need to worry about unemployment when the mechanical cotton-picker or the computer was developed. Even if this long-run view is held, the lives of people in the short-run are important. A scientist may not feel inclined to say that he will not investigate a field, because his findings may have short-run bad consequences. He may, however, feel responsible for warning policy-makers of his findings in order that they can plan compensating lines of action. The present concern of bio-medical scientists with the ethical consequences of genetic discovery and with problems of genetic engineering is a clear determination not to be caught unawares in the way nuclear physicists were unprepared for the military application of their work.

The unpredictability of consequences as well as the prevalence of unintended consequences is well illustrated in the history of the automobile whose technical and sociological consequences meant vast unforeseen changes in our mode of life; and when we thought we knew them all there came the fresh realization of effects in polluting the atmosphere. Responsibility does not mean that the early inventors are to be held accountable for the total outcome. But it does suggest the need for cooperative scientific responsibility to monitor systematically the applications of science in a continuous, not an intermittent, institutional manner.

There is, of course, a gap between science and social policy that is filled by knowledge other than the findings of the particular scientist and by values and value-judgments in the community. The scientist in making his findings public should recognize this gap and indicate the multipotential effects of his own work on different value-judgments and on different assumptions of other knowledge. He may thereby be able to excuse himself from specific social responsibilities, but if he draws conclusions for policy, he should make clear what other knowledge he is using and what the value stand involved is taken to be.(23) Thus we would expect a believer in the role of heredity in intelligence and its application to race to make points such as the following clear, if he is drawing conclusions for education: 1. How his recommendations for educational policy would have any relation to researches on race in an educational system, the professed ideal of which is the treatment of each individual as an individual person. 2. What the underlying, possibly alternative, ideals of educational apportionment would be. For example, is each child to be educated for a rounded development of his native abilities, or only for what we can predict he will do best, or only for what there is current social or economic need? Or does education involve a conception of scholarship as a capacity of an elite group that is to be given the greatest scope? 3. What life rewards of different educational tracks are to be taken to be, and whether they are occupational success and higher gains or the values of a type of cultivation of the spirit.

In short, the consequences of the hereditarian findings (setting aside the issue of correctness) may vary widely for different conceptions of the nature of the educational process, the aims of education, the supposed rewards of education, and the presumed social consequences of tracking. If the scientist who is recommending changes in policy does not assume a responsibility for clarifying the concomitant bases of that policy, he should hardly be surprised that the worst view may be taken of his purposes in making the recommendations. In critical social controversies lack of clarity often borders on obscurantism, which is a traditional property of ideological uses of knowledge.

The same cautions are relevant for a scientist who is not himself translating his findings into policy recommendations, but whose findings are nevertheless a ready intellectual weapon in an existent social conflict.

Finally, the scientist who becomes an advisor or expert for a particular legislative campaign or some immediate alteration of executive policy has assumed this status voluntarily. The ethics of campaigning, much in flux today, are then relevant. An inquiry by moral philosophers into this area would be fruitful, and illuminating comment can be found in the ancient Stoic philosophers on what must be endured if one decides to enter politics actively. But this at any rate is clear in our present inquiry: a scientist serving in the political context cannot claim that his freedom of scientific inquiry into race and intelligence is being jeopardized if he is attacked for sending his children to private rather than public schools, or if he has not resigned from an all-white club.

In the light of all these considerations, what can be concluded about the judgment that under present conditions research into the problem of race and intelligence is inadvisable? It was suggested that this type of question is a legitimate one for the individual scientist to ask himself about his own future action. Some physicists have refused to work in a whole area in which war uses of their discoveries are at present unavoidable. In the area of race and intelligence, there is less clarity and greater complications than for the physicists. There are dangers of intensified racial discrimination. Chomsky compares such work in research to that of "a psychologist in Hitler's Germany who thought he could show that Jews had a genetically determined tendency toward usury (like squirrels bred to collect too many nuts;"(24) the very raising of the issue of Jews and usury would lend weight to the persecution of Jews.

There is, however, the fact that the question of race and intelligence has long been asked and might best be answered by intensive scientific investigation and criticism. Unfortunately the research itself may not be feasible, because the social conditions under which we live may preclude doing the central studies which involve, for example, control of nutrition, prenatal environment, hormone level as it affects anxiety in mothers, as well as social attitudes. It is also possible that the question of race and intelligence is so inextricable from ideological conflict that it would be impossible to keep it as a purely scientific one for a given period. A further issue is whether the research has any scientific validity. At the present stage of genetic knowledge and of conceptualization of heredity-environment interaction, the formulation of research may be too gross and too speculative for useful work. Generally, the libertarian tradition would be inclined to stress scientific grounds in foregoing a line of research, but a decision based on social consequences cannot be eliminated for individual scientists who find it a regretful outcome of a rational balancing of values. Probably different scientists will reach different personal conclusions.

I have tried to show that the responsibilities of the scientist in the translation of his findings into public policy are great; that the issues are complex; that the route is indirect and involves other knowledge, values, and ideals; that responsibilities extend back to the selection of research areas and the modes of experimentation. However, the scientist's primary responsibilities are scientific and involve clarification of relations. He is not necessarily to be saddled with agitational or activist duties, for how a person partitions his activities is a separate moral question.

Finally, I have not attempted to pass judgment either on the scientific merit of the findings in the controversy over race or intelligence or on the extent of ideological intrusion into the controversy. And I have not attempted to assign blame or responsibility for the present sad state of the controversy. I am inclined to think that considerable responsibility lies in the whole state of the psychological and social sciences and philosophy today—not only in the traditional isolating beliefs of value neutrality, but in the dogmatic habits of schools of thought and their intransigent conflicts. Differences of view, of method, even of beliefs can be enriching if they are examined in a responsible dialogue and scientific interaction. If instead the attitudes and habits of war or the game model of the football field govern the procedures of intrascientific relations, what can scientists expect in the public forum but that youth should absorb this lesson and extend it with corresponding youthful vigor? A framework of civility should begin at home.

NOTES

(1) Walter F. Bodmer and Luigi Luca Cavalli-Sforza, "Intelligence and Race" in *Scientific American*, October 1970, p. 29.

(2) William Shockley, "The Apple-of-God's-Eye-Obsession" in *The Humanist*, January/February 1972, p. 16.

(3) For a concrete situation in which choice would be forced on us, consider legal cases in which the possibility of knowing the truth is rejected because of the illegal means by which it is ascertained. Bernard Botein and Murray A. Gordon, in their book, *The Trial of the Future* (New York: Simon and Schuster, 1963, ch. 2) discuss the likely future situation in which a jury could be dispensed with and hypnosis of the parties or drug-induced revelations substituted for jury verdicts; the choice would be between a ready means of truth and preserving human dignity.

(4) John Stuart Mill, the central apostle of liberty in the liberal tradition, assigns so important a place to the atmosphere of rationality that he declares the principle of liberty to have application only when society has reached a stage in which men settle issues by rational discussion. Mill's *On Liberty*, chapter 2, is the classic defense of freedom of thought and expression.

(5) See, for example, the on-going discussions by geneticists to defer lines of pure research because of their potential hazards; it concerns gene-transplanation accidentally creating drug-resistant germs or new types of cancer causing viruses. See also the researches and publications of the Hastings Center of the Institute of Society, Ethics and the Life Sciences. For an example of the growing sensitivity to moral aspects of psychological experimentation, see the discussion of Stanley Milgram's experiments in the review of his *Obedience to Authority* by Steven Marcus (The New York *Times* Book Review, January 13, 1974, pp. 1-3).

(6) An episode of this type in the field of cancer research is reported in The New York *Times*, Thursday, April 18, 1974, p. 20.

(7) P.W. Medawar, in a review of H.J. Eysenck's *The Inequality of Man*, concludes: "One of the most important things laboratory research teaches us is how very often we are quite mistaken, but this is a disability of which Eysenck betrays very little awareness." (*New Statesman*, January 11, 1973).

(8) Arthur R. Jensen, "How Much Can We Boost I.Q. and Scholastic Achievement?" *Harvard Educational Review*, Vol. 39, No. 1, Winter, 1969, pp. 1-123.

(9) A story told about John Dewey's reaction to intelligence testing may still be salutory today. Attending a conference on the question at Teachers College, Columbia, he was asked by Thorndike to comment. He said intelligence testing reminded him of the way in which people used to weigh hogs in Vermont: a strong rail was balanced carefully on a fence; the hog was placed on one end of the rail and stones on the other end. The hog's weight was then determined by guessing the weight of the stones! For a clear presentation of methodological issues in intelligence testing, see N.J. Block and Gerald Dworkin, "I.Q.: Heritability and Inequality, Part I," in *Philosophy and Public Affairs* Vol. 3, Number 4 (Summer 1974).

(10) For a study of the genetic and physical factors involved, cf. essays by W.F. Bodmer, John Hambley, and Steven Rose, in *Race and Intelligence*, edited by Richardson and Spears (Baltimore: Penguin Books, 1972).

(11) For a critique of the twins studies, cf. Leon Kamin, "Heredity, Intelligence, Politics and Psychology," paper delivered at the Annual Meeting, Eastern Psychological Association, Spring, 1973.

(12) The cultural critique was raised in the 1930's by the research of Franz

Boas and Otto Klineberg. It is recently supplemented by a new look at black cultural forms as specializations of their own rather than impoverished derivatives of white cultural forms; see, for example in the case of language, W. Labov's study of the independent character of black children's speech, in his "Logic of Non-standard English" in Frederick Williams, ed., *Language and Poverty* (Chicago: Markham Pub. Co., 1970).

(13) Cf. *The New Assault on Equality: I.Q. and Social Stratification*, edited by Alan Gartner, Colin Greer, and Frank Riessman (New York: Perennial Library, Harper and Row, 1974).

(14) Jensen has been criticized as adding many weak strands to reach a strong indicative conclusion. As for emphasis, J. McV. Hunt decries the initial position given (in Jensen's first sentence) to the view that "Compensatory education has been tried and it apparently has failed." It is, says Hunt, a dangerous half-truth: "I find it hard to forgive Professor Jensen for that half-truth placed out of context for dramatic effect at the beginning of his paper." ("Has Compensatory Education Failed? Has It Been Attempted?" reprinted in *Environment, Heredity, and Intelligence, Harvard Educational Review*, 1969, pp. 148, 149).

(15) The criticism of inadequate responsibility in the controversy is a wide one. For example, the Eastern Psychological Association voted the following resolution at its annual business meeting at the 44th Annual Meeting: "The EPA wishes to reiterate its long-standing commitment to a policy of strict adherence to scientific principles in research. Because this is especially important for research on social issues, the EPA censures the use of inconclusive evidence concerning race and I.Q." (reported in *American Psychologist*, Sept. 1973, p. 767). Cf. also Sandra Scarr-Salapatek's conclusion on this question in her review of works by Jensen, Eysenck and Herrnstein (*Science*, December 17, 1971, p. 1228) "And to assert, despite the absence of evidence, and in the present social climate, that a particular race is genetically disfavored in intelligence is to scream 'FIRE! . . . I think' in a crowded theatre."

(16) Of the articles that initially fired the controversy, Jensen's appeared in the *Harvard Educational Review*, which may be regarded as a semi-public medium; Richard Herrnstein's "I.Q." appeared in the *Atlantic Monthly* (Sept. 1971, 43-64).

(17) Reported in The New York *Times*, December 16, 1973, p. 1. See also the discussion in The New York *Times*, editorial section, Sunday December 23, 1973, p. 5.

(18) In correspondence with the chairman of SPSSI's Fact-Finding Commission on the Suppression of the Academic and Scientific Freedom of Hereditarian and Behavioral Researchers and Teachers.

(19) The sociology of school conflicts is only beginning to be studied more intensively. It relates to larger theories interpreting the history of science—especially on the question of continuity and discontinuity in development. Some of these issues emerge in Thomas Kuhn's well-known study, *The Structure of Scientific Revolutions* (University of Chicago Press, 1962).

(20) Quoted in Norman Daniels, "The Smart White Man's Burden" in *Harper's*, October 1973, p. 25.

(21) That is, of course, if we think the question worth arguing. A somewhat different approach is taken by Bayard Rustin, who condemns the harassment of Shockley as giving undue publicity to an unimportant issue: "The economic and

social environment of black people was altered quite dramatically during the 1960s and black people, as a result, have achieved unprecedented progress. For blacks, the challenge is to continue to dedicate ourselves to winning important, mass gains, not to creating martyrs out of proponents of irrelevancy." (*The New York Teacher*, January 20, 1974, p. 20).

(22) In an article on "Reducing the Heredity-Environment Uncertainty" (reprinted in *Environment, Heredity, and Intelligence, Harvard Educational Review*, 1969), Jensen replies to several critics of his original article. In a section on Social and Educational Policy he says: "I am not a social or educational philosopher and I am sure that neither I nor anyone else at present has thought through all the policy implications of my article." (p. 239) This kind of warning might well have had a central place in the original article, but it would also require amplification about what is involved in thinking through policy implications.

(23) Medawar points out in the review of Eysenck cited in (7) that usually when we think of inherited differences in intelligence we take refuge in the area of equality of opportunity. (Presumably this is on meritocratic assumptions.) But, he says, we could equally call for greater opportunities when compassion and a sense of justice step in, as we do in the case of children with genetic defects who have greater medical needs.

(24) Noam Chomsky, "The Fallacy of Richard Herrnstein's I.Q.," reprinted in *The New Assault on Equality*, supra note 13, p. 98. Chomsky assumes that only a racist society would raise the question of correlating intelligence with race. However, we note that it is the same society that has raised the question of the ideological character of such in inquiry.

RACE, REIFICATION, AND RESPONSIBILITY

Robert Cancro

If there is any principle of the Constitution that more imperatively calls for attachment than any other, it is the principle of free thought—not free thought for those who agree with us but freedom for the thought that we hate.

—Oliver Wendell Holmes

The events that led to publication of this book are not only interesting in their own right, but help to identify some of the nonrational factors which seem critical to this observer. During the winter of 1971-1972 Ellis B. Page drew up a two-part document on behavior and heredity. The first section described very briefly certain threats to the freedom of scientific inquiry within the academic community which troubled him. There followed a five-part resolution defending the right of scientists to do research into the biologic as well as the nonbiologic basis of human behavior. It supported the importance of such a line of inquiry as a complement to the environmental approach. It also deplored the lack of weight given to the role of heredity in a number of disciplines. Some fifty scientists who shared Doctor Page's concerns to varying degrees independently signed this document which appeared in the Comment section of the *American Psychologist* in July of 1972.

I shall personalize this presentation by only describing a representative selection of subsequent events in which I was involved. While this may be of limited generalizability, it is still of value since it does represent first-hand experience. Several friends to whom the resolution was shown warned me not to sign it since it could lead to my becoming identified with racism in America. This seemed a rather remarkable fear since my liberal credentials were in order and the Resolution made no specific mention of group or even for that matter individual differences in human behavior. One much respected colleague refused to sign since he feared such an act would make him the target of radical students at his university. He subsequently congratulated me privately on my courage. For the sake of historical accuracy it should be noted that my signature was more a reflection of naivete than courage. The signing of this Resoultion lead to the receipt of a number of letters ranging from critical to threatening. These included a demand on the President of my university that I be fired for having signed this document. Late in the year, a letter arrived from SPSSI informing me as a signer that a Commission of the Renewed Assault on Equality had been formed which wanted to study the social meaning of this Resolution. I inferred from the rather staccato-like listing of their five questions, particularly numbers three to five, that there was a hostile tone to their inquiry. This is not to say that the letter implied any hostility to the Resolution, but rather to emphasize that this was my inference. This distinction between inference and implication which has been lost in the third edition of Webster's remains important and will be developed somewhat later. A lengthy correspondence ensued in which positions were clarified and arguments aired. My initial ruffled feelings were sorted out and I felt much more comfortable about the intent if not the tact of this unfortunately named Commission. It was difficult not to infer bias or prejudgment from the name chosen by the Commission and even more difficult to grasp the therapeutic benefit of a name change. Nevertheless, the emphasis of their concern seemed to me to change dramatically over time. Initially, the Commission knew "of *no evidence* (emphasis mine) that there is academic or scientific suppression and censure. . . ." This rather strong and clear statement subsequently changed to *systematic suppression* (emphasis mine) of a hereditarian viewpoint. Clearly, there is a fundamental difference between these two statements. Correspondence with the Commission revealed it had inferred from the Resolution that the signers were suggesting the existence of an effort at systematic suppression of behavior genetics or what might be termed more simply a plot. There were repeated references to the implications of the Resolution. It would be more fair to speak of their inferences rather than its implications. Speaking for this signer, I never thought at any level—conscious or otherwise—that there was a plot in the academic community to suppress the field of behavior genetics. A wish perhaps, a plot certainly not. The insistence that the Resolution refers to a concerted effort at suppression of genetics is at best a misleading distortion.

There has been rampant irresponsibility within the academic community on both sides of the position supported in this Resolution. The defenders and the attackers have been quick to label each other with a variety of mindless epithets. Having spent some of my youth being denounced as a communist it is perhaps only in the nature of balance that I spend some of my middle years being denounced as a fascist. Many attackers of the Resolution claimed it was a thinly disguised effort to attribute genetic deficit as the cause of black-white IQ score differences. The most generous comment I can make about this inference is that it is of a high order and far removed from the data of the Resolution. A less generous analysis is that the accusation is a deliberate lie intended to politicize the situation and prevent, thereby, any rational debate. The SDS misrepresented the Resolution in some of its "literature" as an advertisement. Perhaps it is not a surprise that the SDS is so careless about the relationship between facts and language. Any organization that includes democratic in its name despite the antidemocratic nature of its activities is likely to be careless about other words as well. What is more difficult to understand is the behavior of colleagues who accept SDS statements as de fide truths requiring no further substantiation. A recent example of this curious trust is a letter to the signers from a member of the scientific community demanding that we withdraw our signatures and repeating the SDS lie that the Resolution was an advertisement. Needless to add, the writer of this letter has no training or competence as a geneticist. He is not even a behavioral scientist.

Many of the attacks on individual scientists which include physical assaults as well as some of the attacks on the field of behavior genetics are more than intemperate. They are nonrational. Ringing denunciations of 19th century eugenics are made side by side with comments about the limitations of contemporary behavior genetics from which I infer the suggestion of a relationship. Perhaps we should demand that our sister science disassociate itself from its historical past. This seems grossly unfair. It is true the field of behavior genetics has a historical relationship to the eugenics movement but it is also true that the field of chemistry has a similar relationship to alchemy. Those of us who are presently or were formerly in academic psychology need not look too far into our own past to find some reasonably absurd precedents which are better forgotten. A recognition of past errors should only make us cautious but not lead to the scourging of present day science and scientists.

This Resolution, intent of which was to defend publicly the right of scientists to pursue the investigation of the role of biologic factors in human behavior, was labeled racist by the American Anthropological Association. It is my understanding that their assessment of the scientific merit of the controversial Resolution was done in a remarkably scientific fashion, i.e., a vote. At least this is a quantitative measure and as such may represent a step forward for them. It is my hope that the Behavior Genetics Association will not retaliate by voting on the genetic competence of the American Anthropological Association. Clearly, there has been such an extraordinary degree of misrepresentation and misunderstanding as to require a psychologic assessment of the origins of this nonrationality.

Again speaking in the first person, I can empathize with the fears and concerns of individuals who dread the potential misuse of scientific information to support racist policies. Frankly, I do not believe racists are influenced by scientific findings one way or the other. To me racism is a type of delusion. It is a fixed, false belief which is not amenable to change through scientific evidence and/or human experience. I also doubt that scientific findings can produce racists. Nevertheless, there is a legitimate concern—shared by this writer—that scientific studies may be used to justify social policies which are destructive or at the very least injurious to certain groups. One approach to the protection of the politically weaker groups in our society is to prevent any scientific investigation that may lead to such an undesirable outcome. Responsible people may argue that there are certain lines of inquiry which should not be followed because the potential risks far outweigh the potential benefits. This is certainly true in situations in which the individual subject is exposed to personal risk with the hope being that another individual will derive benefit at some future time. There is a second concern which weighs very heavily on my own conscience. It is the freedom—particularly but not exclusively for the scholar—to think ideas which deviate from the current norm. As corollaries to this freedom to think, there are the freedoms to study, to discuss, to debate, and the responsibility to modify through the corrective rational feedback from those around you. Since I value all of these very highly, it is necessary for me to find a compromise when and if they come into conflict. However, I feel a very powerful case can be made that the conflict is more imagined than real.

The issues surrounding the determination of human behavior including individual and group differences are charged by an excess of affect. This leads to the inundation of the cognitive processes and to fundamental blunders. An example is that respected colleagues constantly use the word "versus" when speaking of genetics and the environment. This is a basic error which would not be made by a first-year graduate student in biology whose adrenals were in a resting state. Setting aside for the moment the many emotional considerations which have led to nonrational positions, we can identify two historical trends—and perhaps even forces—operating within science that contribute to much of the confusion in the current controversy. I submit that as in most controversies, there is more confusion and ignorance than substance.

Descartes' distinction between the mind and body has been one of the most productive and useful divisions in the history of Western thought—both politically and scientifically. Politically, it was a most fortuitous distinction since it removed the study of the body from church-imposed restraints. By giving unto the church the mind and unto natural science the body, he achieved a working truce between a powerful and a fledgling political force. While the disparity in strength between church and science has shifted, we should not deceive ourselves into believing that free scientific inquiry has the power, for example, of the military-industrial complex. Science and the freedom of thoughtful inquiry which underpins it, is a large but nevertheless delicate entity whose continued existence can easily be threatened. The perversions of science to the will of the state go on in 1976 and do not need to be documented further. The lesson to be drawn from these experiences is that when science and politics mix, it is science that suffers.

Scientifically, the mind-body distinction has been extraordinarily productive. It has freed neurophysiologists to look at molecular function without having to be immediately concerned with molar behavior. It has led to useful differential diagnostic activities in the clinic, including different forms of treatment for disorders which are "organic" as opposed to "functional." In summary, the dualism of Descartes was wonderfully useful as a concept but as with all formulations it has inherent limitations and disadvantages. Any cognitive order that we impose on data excludes certain other useful formulations. Dualism has been productive and now the time has come to go beyond it. We must develop new constructs which synthesize the positions derived from the arbitrary dichotomy rather than to build further upon them. All human behavior is inextricably interwoven with the function of the central nervous system and can only be fully comprehended through the inclusions of an understanding of it. Our task is to unite these different disciplinary insights into a comprehensive theory rather than to restrict ourselves exclusively to any single vantage point. An exclusively biologic explanation of human behavior is reductionistic and suffers from a second problem which we shall delineate momentarily. On the other hand, to ignore the biologic functioning of the nervous system is to be naive and simplistic.

Theoretical formulations including the mind-body distinction which are comfortable ways to thinking about data are treated as if they are Platonic truths. This tendency to reify, poisons much of contemporary science and leads to violent partisan quarrels. Science is not the pursuit of absolute truth but rather the pursuit of useful ways of thinking about data. A tendency to treat the mind-body distinction as a real one has even led to different schools of thought concerning human behavior which are often dubbed biologic versus psychologic. Unfortunately, the use of the word "versus" in this context accurately depicts the reality.

It may be helpful at this juncture to review briefly the actual position of behavior genetics in contrast to the reified version. An important theoretical contribution of the field has been its identification of the artificiality of the gene-environment distinction and its insistence on recognizing genetic and nongenetic factors as essential codeterminants of human behavior. Behavior genetics clearly states that the distinction between the environment and genes is arbitrary although useful. No genotype operates without an evoking environment and no environment can evoke without the presence of the genotype. Dobzhansky (1964) has clearly made the point that both are necessary and, therefore, both are equally important in the case of the *individual* phenotype. In a real sense the environment determines which of the large number of genes making up the individual's endowment will become the effective or operational genotype by selective enhancement and suppression. An excellent illustration of the limitation of semantic conveniences in approximating biologic realities can be drawn from a classic experiment in behavior genetics (Zamenhof, et al. 1971). A normal group of female rats when fed protein-deficient diets will produce offspring who show fewer and smaller brain cells. The brains of these young rats do not respond to a normal protein diet. Man's cognitive apparatus interprets this as an environmental effect which is uncorrectable by further dietary manipulation. Nevertheless, these "uncorrected" rats also produce offspring with a reduction in number and size of brain cells. It takes several generations for this so-called environmental effect to disappear. Are we to conclude that an environmental effect was transmitted in reproduction? It is wiser to recognize the limitations of our cognitive apparatus rather than to insist that the laws of nature must approximate in every particular our disciplinary ways of thinking. It is vital that at the very least the limited goal of laying to rest the false issue of heredity versus environment be achieved.

It is generally believed by behavior geneticists that the gene even when activated does not determine the phenotype in an inevitable and totally predetermined fashion. There is a range of possible phenotypic outcomes of technically phenoptions inherent in any given genotype. Genetically identical individuals will show different outcomes if exposed to different evoking environments and genetically different individuals will show the identical phenotype when exposed to different environments. In other words, individuals can be isogenic without being isophenic or can be isophenic without being isogenic. Clearly, this insight further complicates matters and serves to make simplistic statements about individuals—be they genetic or environmental in content—less tenable.

It may be useful to shift the focus now from the individual to the group. It is obviously possible to take a population of individuals and measure through the appropriate mathematical techniques a proportion of within group variance attributable to genetic or nongenetic factors. The estimate would be valid for this population at this given time in its history if the trait in fact follows the mathematical assumptions of the method for deriving the estimate. It would not be predictive of a genetically different population or even of the same population in a different environment. Even if the estimate of heritability were to be approximately the same in a second population, this finding would only have limited usefulness in terms of speculating as to the origin of any between group differences. All of this is well known and only requires repetition because it appears to be equally well forgotten.

The politicized climate is such that the contributions of behavior genetics to an understanding of individual and group differences can not be stated in general terms but must be specifically applied to and defended in the IQ debate. The insights of genetics concerning IQ are real but do not readily translate into policy. Nevertheless, it has become a popular pastime in recent years to attack IQ although it is beyond question the best measure ever developed by psychology. It seems absurd to have to recognize that this single measure has multiple limitations. What is remarkable and deserving of careful scrutiny is its predictive power. There are many types of cognitive ability worthy of the designation intelligence. IQ does not purport to measure all of these. It does, however, measure in a reasonably reliable way certain abilities—particularly abstract categorical thinking—which are predictive of certain performances including socioeconomic in our culture. Considering how many other factors including motivation go into the determination of one's socioeconomic destiny, it is amazing that the IQ score predicts at all. We must ask if IQ is diagnosing a social problem which intellectuals are reluctant to face, a problem which is not racial at all but which some intellectuals prefer for self-serving reasons to identify as such. Oriental-white and black-white differences may highlight the problem but they do not, in my judgment, comprise it. This speculative hypothesis will be developed later as a third trend that contributes to the confusion.

The most conservative analysis of within group differences attributes 80% of the IQ variance to genetic factors and only 20% to the environment. Yet, according to this conservative analysis of relative contributions, a one S.D. difference in total environmental effect would equal 6.7 IQ points ($\sqrt{\text{variance} \times .20} = \sqrt{15^2 \times .20}$). As much as 6.7 IQ points of difference between genetically identical white individuals could theoretically be explained by one S.D. of total (direct & indirect) environmental effects. (The exclusive use of white identical twin data would yield a S.D. of 4.74 as the measure of total environmental effect.) Furthermore, there is no single genotype for a particular IQ score or range. The identical phenotype, i.e., score, can be achieved by the same or different genotypes. If the unwise leap were made directly from within to between group differences using the identical approaches, the assumption of one standard deviation of inferiority in the black environment could explain up to 45% of the group IQ difference.

This illustration is not meant to be offered as evidence nor does it imply that a fixed percentage of the between group difference in black-white IQ scores is attributable to group genetic differences but rather to illustrate how reflecting on the genetics dispassionately forces us to extremely cautious conclusions. One would have to be rash indeed to suggest drastic measures when a significant amount of the group difference can be accounted for on the basis of nongentic differences making the most conservative genetic assumptions. More importantly, behavior geneticists generally believe that any large breeding population—and ultimately that is what defines a race—is equipotential with all other large breeding populations. There are no known genetic differences between Orientals, whites, and blacks which are critical to the best of our present knowledge. The only interesting differences between races are in the relative frequencies of particular genes. (There are a few genes which are unique or almost unique to particular racial groups but we believe these to be biologically unimportant.) Gene frequency is determined by the environment in which that breeding population has lived. In a very real sense, every race represents an independent gene-environment experiment. Differences in gene frequency do not produce inferiority nor superiority. More importantly, although there are differences between individuals and between groups genetically, the outcome of that gene-environment interaction can be influenced through manipulation of the environment. If our goal is to achieve identical phenotypes, it is theoretically possible within relatively broad boundaries to take different genotypes and place them in different environments and produce the same phenotype. Fortunately, mankind does not possess the knowledge to manipulate the environment in this fashion as yet. Certainly, it does not have the wisdom to exercise that power should it be developed.

It is hard to believe that there are serious scientists who deny the importance of genetic factors in human behavior. Are we to believe that behavior genetics is a fraud and that humans are the only form of life impervious to genetic forces? Genes operate in human affairs. The more we understand their influence, the safer we are from ignorance and/or malice. There is no doubt if we made the environment identical that certain individual and even group differences would still exist. Only now they would be totally genetic in origin. To wipe out individual and/or group differences while maintaining genotypic diversity would require the ability to identify the genotype, the specific evoking environment, and the necessary timing of the interaction that would create this uniform, Orwellian nonperson. Only then would it be true to say that all men are equal!

It is a far cry from these limited insights of behavior genetics to social policy. I, for one, would be hard pressed to derive specific social policy from my knowledge of the importance of genetic factors in human behavior. Clearly, it is the height of intellectual arrogance to suggest that the environment of the poor should be improved so that their children will have higher IQ's. The moral obligation that a wealthy country such as ours has to define a standard beneath which a citizen is not forced to live is clear, compelling, and independent of IQ test scores. The equalization of opportunities and the ending of discriminatory practices is a matter of justice and not knowing what to do with a stamped addressed envelope. It is certain that genetic diversity must be maintained if man is to have a biologic future. It is almost equally certain that environmental diversity and ease of movement between environments are important if the greatest number of people are to achieve their richest potential. One limited policy experiment that derives from these considerations would be the maximization of educational diversity. There may be more than political wisdom in the creation of school curricula for black ghetto children in Newark which emphasize Swahili, African history, and alternative learning atmospheres. It would be ironic if a former playwright showed more educational creativity and genetic sophistication than the professional educators.

A speculative hypothesis derived from the predictive value of IQ will now be presented as a third—probably unconscious—factor fueling the controversy and one to which intellectuals are particularly susceptible. While I do not have the wisdom to translate this speculation if valid into policy, it may be useful to discuss it. In a technologically advanced society, the IQ test is an excellent measure of those abilities that are valued and rewarded by the society. The future in most technologically advanced countries belongs to those with an IQ above 115. This group initially contains 16% of the population for whites. When we eliminate those who choose to drop out or who lack the desire to compete or are hindered in their progress through sexual and other forms of discrimination we wind up with a very small subset of the population that represents the elite class in the technologic society—the intellectuals. The intellectual elite class would be no larger than 10% and probably less of the white population. It is this very class that tries to discredit the IQ test and thereby, obscure its diagnostic implications. The implications may be more for nonintellectuals than for nonwhites. It is hard for me to see the significance of the exact percentage of intellectuals for Orientals and blacks. If the vast majority of citizens perform tasks that are relatively devoid of prestige then you have a social system that is inherently unjust and, therefore, unstable. It is necessary to reshape and alter our society so as to reward and more importantly to value a variety of humans and their discrete activities. If our social organization does not reflect and respect the diversity of humanity, then it will not come as a surprise that its institutions particularly educational ones try to homogenize the young into a single useful product.

RESPONSIBILITY

The final issue is the one of responsibility. The prime responsibility of the academic community is to guard the precious fire of free inquiry. The university remains one of the few places where people can safely think unpopular thoughts. Tenure was once meant to protect academics from the possible negative consequences of such activities. Today, the dangers are more subtle and more from within the community. The university and its individual members cannot tolerate nor permit the silencing of heresy. When we start to throttle a man's thoughts we often end up by throttling him as well. This is not to say that there are not other important values in the university but only that open and free inquiry is its highest value.

Every individual academic has additional responsibilities as a citizen. These include participation in the political process of policy making. Not every citizen will chose to involve him- or herself as deeply in these matters as do others. This is as it should be. Nevertheless, if an individual's research lends itself to ready misunderstanding, that person has a greater moral responsibility to society than does a colleague whose work is more esoteric. Even if some colleagues are remiss in meeting this responsibility, it would not justify undermining that freedom of thought which is guaranteed in the academic community.

The currently popular euphemism for control is accountability. It is a cleverly selected word since it suggests as its alternative unaccountability, which is morally reprehensible. There are two important questions that must be raised about accountability. These are for what and to whom. The danger in being accountable for one's thoughts lies in the constraints it puts on the freedom to think. While this freedom is essential in the university, it is extremely important in other human arenas as well. There can be no meaningful freedom of speech in the absence of freedom of thought. The even greater threat lies in accountability to any political group however selected which seizes the power to decide what is correct and responsible thinking. Accountability does exist in the academic community and in particular in its scientific subcommunity. This accountability derives from the openness of academic inquiry including replication. The frauds are discovered and revealed. This process is, however, not a substitute for the ultimate accountability which is to one's own conscience. This internal accountability holds equally for the scientist as it does for the war resister.

The freedom to think and to disseminate deviant ideas is extraordinarily fragile. At the individual level we tend to reject not only the idea but the person. When George Lincoln Rockwell was stoned in Union Square Park by American Jewish veterans, I was delighted. At an intellectual level their action was repudiated but at a more powerful level it was valued. Years later when Rockwell was assassinated it came as no great surprise to discover that my internal split between constitutional law and affective retribution had not been healed. I was momentarily glad that he was dead. This ugly pleasure in the destruction of repugnant ideas and their originators is a weakness not restricted exclusively to this writer. This difficulty in allowing, let alone protecting, deviant thought is ubiquitous.

At the group level, even a casual study of history reveals that politics and free inquiry coexist only at a distance. The tragic lesson of Nazi Germany is not that there were scientists who cooperated with the state, but rather that the destruction of the free and open—and therefore self-correcting—inquiry of the academic community was the necessary condition to allow this cooperation to continue in an undisputed manner. When political groups control the freedom of some people to think, the rest of the population is in immediate peril. No group however appointed and/or annointed can decide what are the correct ideas. It is this thought coercion whether it comes from the political left, center, or right that is the ultimate form of fascism and a malignancy that we must destroy or which will consume us all.

REFERENCES

Dobzhansky, T. *Heredity and the Nature of Man.* New York: New American Library, 1964.

Page, E.B. "Resolution on Behavior and Heredity." *American Psychologist,* 27: 660-661, 1972.

Zamenhof, S., van Martins, E., and Gauel, L. "DNA (cell number) in Neonatal Brain: Second Generation (F_2) Alteration by Maternal (F_0) Dietary Protein Restriction." *Science,* 172: 850-851, 1971.

BEHAVIORAL SCIENCE AND SOCIETY: THE NATURE– THE NATURE-NURTURE CONTROVERSY AS A PARADIGM

David Layzer

Arthur Jensen's now-famous article in the *Harvard Educational Review* appeared early in 1969, and by August of that year—according to an article in the New York *Times Magazine* that contributed in some measure to the phenomenon it described—"Jensenism" had become a household word. In my own household, however, Jensen's doctrines received scant attention until two years later, when the September 1971 issue of *The Atlantic Monthly* showed up on our coffee table. The cover was provocative. Beneath the massive black letters I.Q. were displayed a fragment of an I.Q. test and a list of seven questions:

> Which principally determines intelligence? 1. heredity 2. environment. Can schooling raise your child's I.Q.? Do bright parents have brighter children? Do dull parents have duller children? Is there an emerging ruling class of the intelligent? Is there a permanent lower class of the unintelligent? Does I.Q. testing discriminate against the poor and the black?

As I read these questions, it crossed my mind that the article, by Richard J. Herrnstein, professor of psychology at Harvard, might stimulate lively discussion among Harvard undergraduates returning to or newly arriving at Cambridge after the long summer vacation.

In the actuality, it also stimulated a variety of less rational responses—demonstrations, denunciations, heckling, and even personal harassment. These raised a new set of issues. Professors argued that unrestricted freedom of inquiry and expression in an atmosphere of civility and mutual respect are essential to the functioning of an academic community. This view enjoyed widespread but qualified support among the students. Although nearly all of them deplored the tactics of harassment, few were satisfied with the faculty's response to the substantive issues raised by Professor Herrnstein's article. They were prepared to accept the proposition that in a university one must fight ideas with ideas, but they could not help noticing that, with few exceptions, those of us who were presumably best qualified to discuss the ideas in question preferred to maintain a discreet silence. This circumstance tended to weaken the argument that truth is forged in the smithy of uninhibited debate. It seemed to me that the least I could do was to read the article and prepare myself to discuss it intelligently.

The article begins with an enthusiastic account of mental tests and their development. ("The measurement of intelligence is psychology's most telling accomplishment to date.") There follows a popular summary of the arguments concerning the heritability of I.Q. that led Jensen and other experts to conclude that "the genetic factor is worth about eighty percent and ... only twenty percent is left to everything else." Using this conclusion as his major premise, Professor Herrnstein then argues that if artificial barriers to social and economic mobility are eliminated, there will emerge a hereditary meritocracy, "a society sharply graduated, with ever greater innate separation between the top and the bottom, and ever more uniformity within families as far as inherited abilities are concerned." Nor is this merely a utopian projection: "The data on I.Q. and social-class differences show that we have been living an inherited stratification of our society for some time." Rank, wealth and prestige are thus seen to be the outward and visible signs of inward and inherited qualities. This thesis is of course scarcely new, but the arguments offered in its support have changed: formerly they were religious, now they are scientific. This change is important. Science has inherited the authority formerly vested in religion; but the pronouncements of science, unlike those of religion, are vulnerable to rational criticism and empirical refutation. Now, the principal scientific arguments offered by Jensen and Herrnstein concern the heritability of I.Q. I was therefore especially interested in the part of Professor Herrnstein's article that dealt with this topic; and as I read, I grew more and more puzzled. Although I knew nothing about I.Q. tests or the statistical techniques used to estimate heritability, I had dealt with the problem of drawing theoretical inferences from experimental data in diverse scientific contexts. From Professor Herrnstein's necessarily simplified and abbreviated account, I found it impossible to imagine how Jensen and the experts whose work he had drawn on could have reached their conclusions from the data they had to work with.

Seeking enlightenment, I turned to Jensen's own article and to the commentaries and criticisms that had been published in subsequent issues of the *Harvard Educational Review*. This left me more puzzled than ever, for neither Jensen nor his critics seemed to have addressed themselves in a consequential way to what seemed to me to be the central scientific issue: the validity and significance of I.Q. heritability estimates. My curiosity was now aroused. I suspected that there might be something wrong with the scientific base on which Jensen and Herrnstein had erected their educational and socio-political theories, and I thought it might be useful and interesting to follow up this hunch. The process of doing so has proved instructive in many ways, and has led me to certain broad conclusions concerning the role of science in society and the social responsibility of the scientist.

I propose to elaborate three main theses, drawing heavily on the I.Q. controversy to illustrate my arguments. 1. The ethical status of a given piece of scientific research can be usefully discussed only after its scientific validity has been accurately assessed. 2. The criteria for assessing the validity of scientific conclusions are—or should be—the same in the physical, biological and social sciences. 3. Scientists should assume responsibility for making scientific value judgments and for accurately labeling the products of their research. The scientific community as a whole must enforce these responsibilities if it is to play a constructive role in shaping public policy.

SCIENTIFIC VALIDITY AND ETHICAL VALUE

Racial differences in "native intelligence." Jensen, Eysenck, Shockley, and others have argued that reported differences between the average scores of U.S. blacks and whites on "intelligence" tests are probably caused mainly by differences in "native intelligence." The principal empirical argument offered in support of this conclusion is that the differences persist when parental income and socioeconomic status are controlled for and when test items judged to be "culture-bound" are eliminated in favor of test items thought to assess "pure reasoning ability." This argument has a plausible ring; but let us examine it more closely.

U.S. blacks and whites constitute two genetically and environmentally distinct populations. The *known* genetic differences between these populations relate to the *frequencies* of a relatively small number of genes that specify physiological characters such as pigmentation, hair color and form, and the like. There is no direct evidence for systematic differences in the frequencies of genes specifying behavioral characters, nor is there any compelling theoretical reasons for supposing that significant systematic differences in the frequencies of such genes exist.

On the other hand, the two groups unquestionably differ significantly and systematically in environmental factors that are believed by students of human development to exert a strong influence on the development of cognitive abilities. Some of these environmental differences have to do with differences in cultural values, customs, patterns of child rearing, etc. Others have to do with ways in which the two groups interact and with differences between the "psychological climates" they experience. Although such environmental differences are conceded by all students of human development to be important, their effects have not yet been carefully studied. Moreover, the notion that certain tests assess "pure reasoning ability" in a "culture-fair" way is not a scientific finding but an article of faith with no visible means of scientific support.

In these circumstances, it is clear that the Jensen-Eysenck-Shockley "hypothesis" is not really a scientific hypothesis at all: it can neither be refuted nor confirmed. The same is true, of course, of the alternative "hypothesis" that blacks are genetically superior to whites with respect to "native intelligence." It is also clear that until we have eliminated or learned to compensate for the relevant environmental differences between blacks and whites, the question whether blacks and whites differ, on the average, with regard to the genetic factors underlying performance on standardized mental tests, must remain unanswered.

This state of affairs is not uncommon in the physical and biological sciences. The following example, from astronomy, is typical. In the spectra of certain hot stars helium lines are abnormally weak. Naively interpreted, this observation suggests a deficiency of helium in the atmospheres of these stars. But the spectra are also abnormal in other ways, and it could well be that the helium lines are weak for some other reason. Until *all* the factors significantly affecting the strength of the spectral lines can be adequately assessed, astronomers must suspend judgment. In this example the question that cannot be answered concerns a well-defined and theoretically significant quantity: the relative abundance of helium in a stellar atmosphere. The question of racial differences in "native intelligence" concerns a commonsense concept that has no precisely defined scientific counterpart. Thus the Jensen-Eysenck-Shockley "hypothesis" is doubly unscientific: it is an untestable assertion about a scientifically undefined concept.

How has such a blatant non-hypothesis come to be accepted so widely as a genuine scientific finding? The answer to this question goes to the heart of the psychometric-statistical approach to the study of human behavioral traits. Psychometricians bypass the difficult questions of what human intelligence is and how it develops by defining it to be a quality or set of qualities measured by standardized tests. They bypass the difficult and important question of how variations in environmental factors affect the development of human intelligence by decreeing that only the factors they have been able to quantify and measure are relevant. Viewed in the context of developmental biology and psychology, these "simplifications" make no sense, but to recognize their absurdity would be to recognize a fundamental inadequacy of the psychometric-statistical approach to the study of human behavioral traits. It is not, perhaps, surprising that practitioners of this approach have been unwilling to take this step.

Heritability. The study of genetic diversity *within* a given population seems at first sight to present less serious methodological problems than that of systematic genetic differences *between* populations. The traditional measure of genetic variation within a given population is the *heritability.* There are two distinct kinds of heritability. The broad heritability of a trait (in a given population under a given range of environmental conditions) is the fraction of its variation (measured by the variance, or mean square deviation) that would persist if all the individual environments were replaced by a single standard environment. Animal geneticists estimate this kind of heritability by analyzing measurements carried out on genetically identical animals (usually monozygotic twins) reared in randomly and independently selected environments. Such measurements yield estimates of the purely environmental contribution to the variation of the trait. Now, the genetic part of the variation of a trait can itself be split up into two components: a part that results from differences that are passed on from parent to offspring, and a part that is non-inheritable because it depends not on individual genes but on *combinations* of genes that may be broken up in the process of genetic recombination. Only the first component of genetic variability can be exploited by the animal breeder seeking to improve his stock through artificial selection. This "inheritable" component of the genetic variation figures in the definition of *narrow heritability*. To estimate narrow heritability, animal geneticists analyze measurements of distantly related individuals, typically half sibs. The distinction between broad and narrow heritability corresponds roughly to the distinction between the notions "innate" and "inherited." The narrow heritability is always smaller than the broad heritability. For example, the broad heritability of milk yield in dairy cattle is typically estimated to be .9 while the narrow heritability is about .25.

The distinction between broad and narrow heritability is relevant to the arguments of Jensen and Herrnstein. Thus when Professor Herrnstein writes, "the data on I.Q. and social-class differences show that we have been living with an inherited stratification of our society for some time," he seems to be referring to data that would afford an estimate of narrow heritability. In fact, no such data exist. The heritability estimates quoted by Professor Herrnstein are estimates of *broad* heritability, which have little relevance to the thesis of "inherited stratification."

Jensen's educability argument, on the other hand, rests on estimates of the broad heritability of I.Q. The argument runs as follows: 1. Children who score low on standardized "intelligence" tests exhibit poorly developed abstract reasoning and problem-solving activities. 2. Estimates of the broad heritability of I.Q. indicate that variations in I.Q. among U.S. whites are predominantly genetic in origin. 3. Hence children who score low on standardized "intelligence" tests lack the genetic capacity to develop abstract-reasoning and problem-solving skills. 4. These children should, therefore, be taught by rote.

Now, even if Jensen's implicit premises concerning I.Q. tests (that they measure abstract-reasoning and problem-solving abilities) and heritability estimates (that they are meaningful and valid) are granted, this argument is fallacious, for it tacitly assumes that a trait whose broad heritability is high is insensitive to environmental variations. This is not the case. A trait may have high heritability and yet be extremely sensitive to appropriate environmental variation. Human height is such a trait. It has high heritability, yet appropriate dietary changes can bring about substantial changes in the average height of a population. Behavioral—and especially cognitive—traits are even more susceptible to systematic environmental changes, such as the change from a predominantly illiterate to a predominantly literate society or from a predominantly rural to a predominantly urban society. Jensen's argument on educability makes sense only on the Panglossian assumption that no systematic improvement in education or in other aspects of the environment that may be relevant to the development of cognitive skills is now possible—a strange assumption for a professor of educational psychology. Yet on the basis of that assumption and of a naive and uncritical acceptance of published estimates of I.Q. heritability, he advocates a radical educational innovation that would deprive children who score low on standardized tests of opportunities to develop problem-solving and abstract-reasoning skills.

An equally serious defect of Jensen's argument is his failure to discuss the evidence that contradicts his key assumption. As Karl Popper has emphasized in his discussions of scientific methodology, the experiments and observations that are most relevant to a scientific hypothesis are those producing outcomes that refute the hypothesis. Jensen's hypothesis on educability immediately raises the question: Are there instances in which children who scored low on standardized tests subsequently achieved high scores on similar tests, or gave other evidence of intellectual achievement? Many distinguished educators agree that the answer to this question is yes. If they are right, then Jensen's hypothesis is untenable. My point is not that they *are* right; I am not in a position to make that judgment. What I do assert is that, as a scientist, Jensen ought either to defend his hypothesis or abandon it.

Heritability and meritocracy. Professor Herrnstein's meritocracy argument illustrates a different kind of confusion between the technical and popular meanings of "heritability." As I have already mentioned, Professor Herrnstein relies on estimates of broad heritability in a context that calls for the use of narrow heritability. But the meritocracy argument rests upon an even more basic technical fallacy. Professor Herrnstein argues that in a society of equal opportunity, prestige and socioeconomic status will be determined largely by genetic endowment. This much is true by definition—though the kinds of achievement that a society chooses to reward and the scale of rewards are susceptible to much greater variation than Professor Herrnstein's argument seems to assume. Professor Herrnstein goes on to argue, however, that as socioeconomic status comes to be determined more and more by genetic endowment, there will emerge "a society sharply graduated, with ever greater innate separation between the top and the bottom, and ever more uniformity within families so far as inherited abilities are concerned." This conclusion is a *non sequitur*; the conditions envisaged by Professor Herrnstein actually ensure continuous socioeconomic mobility.

To begin with, the sentence just quoted illustrates the confusion between broad heritability (a high value of which would lead to "innate separation between the top and the bottom" with respect to the trait in question) and narrow heritability (a high value of which would lead to "ever more uniformity within families"). The distinction between the two kinds of heritabilities is far from academic. I have mentioned the example of milk yield in dairy cattle, which is largely determined by innate factors but is not highly inheritable. Generally speaking, the more complex a trait, the larger the disparity we may expect between the values of broad and narrow heritability. Since the biological determinants of socioeconomic status are likely to be as complex as those of milk yield, it would be most surprising if they should turn out to be highly *in*heritable. Nevertheless let us grant Professor Herrnstein his premise and assume that the behavioral traits bearing on socioeconomic status have a high narrow heritability. It still does not follow that the abolition of nonbiological barriers to socioeconomic mobility would cause these traits to run ever more strongly in families. If the narrow heritability of a trait is high, then its values in successive generations will be highly correlated, but the correlation will diminish rapidly with the degree of genetic relationship. Thus the correlation between grandparent and grandchild will be substantially less than that between parent and child, and that between great-grandparent and great-grandchild will be smaller still.

A simple physical example may help to explain why. Consider a gas maintained at fixed temperature and density. A given molecule moves with constant velocity until it collides with another molecule, when both velocities change abruptly. Think of the molecules as individuals in a sexually reproducing population. Matings are represented by collisions; the molecules that go into a given collision represent parents, the molecules that emerge represent off-spring, and the velocity of a molecule represents a heritable trait. To represent assortative mating, we can stipulate that only molecules with closely similar velocities undergo collisions. Now, if two fast "parent" molecules collide, they are likely to produce fast "offspring molecules;" but the random circumstances of the collision introduce a random element into the outcome. Under these conditions, a given molecule executes a random walk in "velocity space" as it goes from collision to collision. After a few collisions, there will be no way of telling whether its initial velocity was high or low. Similarly, in the kind of meritocratic society envisaged by Professor Herrnstein, familial correlations would be extinguished after a few generations.

It is probably true "that we have been living with an inherited stratification of our society for some time," but Professor Herrnstein's interpretation of this observation is precisely the reverse of the correct one. In view of the preceding considerations the inherited factors primarily responsible for the persistence of socioeconomic status through many successive generations *must* be nongenetic. Among these inherited but nongenetic factors are cultural and familial traditions, wealth, social status, and ethnic background.

The preceding discussion illustrates the dangers of over-hasty generalization. Every genuine scientific concept has an exact meaning and a precisely defined domain of application. Valid scientific inference depends upon strict observance of these limitations. The alleged high heritability of the I.Q. implies nothing about the educability of children who score low on I.Q. tests, about the causes of differences between the average scores of blacks and whites on these tests, or about the causes of social and economic stratification. It may, it is true, lend psychological support to certain beliefs about educability or group differences, but the prejudices of scientists are not necessarily more valuable than those of laymen, and the history of science abounds with examples of similar prejudices that have subsequently been shown to be false.

I.Q. as a biological trait. So far I have focused attention on the scientific implications of Jensen's claim that I.Q. has a (broad) heritability of about eighty percent, leaving open the question whether the claim itself is valid. Although, as we have seen, a high value of I.Q. heritability would not have the implications drawn by Jensen and Herrnstein, it would tell us *something* about educability. It would tell us that, *under currently prevailing conditions,* variations in neither home environment, nor child-rearing and educational practices, nor the quality of schools, nor standards of nutrition and health care, contribute substantially to differences in I.Q. test scores. Many of Jensen's critics have argued that I.Q. scores are not as important as they are made out to be, but this argument cannot be pushed too far. Standardized tests do assess basic cognitive skills—reading and arithmetic in particular—that one needs in order to participate effectively in a modern technological society. If variations in I.Q. are largely genetic in origin, there is no reason to expect that functional illiteracy and innumeracy will be substantially diminished by improvements in the living and working conditions of the poor or by efforts to improve the quality of their education along conventional lines. This is Jensen's central message and it rests squarely on a scientific conclusion that has been confirmed by several investigators. All these investigators, however, started from similar assumptions and took similar analytic routes. Let us take a closer look at the method of analysis and its underlying assumptions.

The first major assumption is that I.Q. scores are valid measurements of a biological trait like height or weight. There are many difficulties with this assumption. The most basic is that the definition of I.Q. is purely instrumental. An I.Q. score is nothing more or less than a conventionally adjusted outcome of an arbitrary test. It is a "measurement" made with a black box. Such "measurements" have no scientific value, for they lack a theoretical framework—something that specifies what the measurement is intended to measure and that enables the scientist to devise distinct and independent ways of measuring it. Lacking a theoretical framework, one has no way of knowing what it is that one is "measuring" or to what extent the "measurement" is affected by systematic errors.

A related difficulty with I.Q. scores is that, though numerical, they are non-quantitative. The I.Q. score is merely a conventional indicator of rank order. But the concept of heritability is meaningful only for quantitative biological characters. From a logical point of view, the phrase "heritability of I.Q." is nonsense, on a par with "factorizability of sunsets." This difficulty will, I think, seem self-evident to anyone who has extensive practical experience in the natural sciences, for such experience quickly teaches one that theories and technical concepts apply only to specific and precisely defined classes of data. In this respect scientific language differs from literary language, where words and phrases acquire new meanings through usage. In scientific language, terms derive their meanings from a comparatively small number of fundamental laws and definitions. From time to time new laws are discovered and new definitions are introduced, but at any given moment the class of meaningful scientific constructs is precisely defined. "Heritability of I.Q." is not now a member of that class.

Numerical indices that lack a theoretical basis may be dangerously misleading. Let me illustrate this point by an example drawn from economics. Personal income is a precisely defined, objectively measurable, numerical quantity, and an important economic datum. Nevertheless, it is not a valid quantitative index of what a sociologist might wish to convey by the term "economic status," because differences in dollar income do not accurately reflect differences in economic status. For example, the difference between an income that allows a family to maintain adequate levels of nutrition and one that does not may be small in absolute terms; indeed, a per capita income that enables one family to buy enough food may be inadequate for another. Thus, if dollar income is used as an index of economic status, qualitative differences of the kind just mentioned are bound to get lost in any quantitative analysis of the data. And if income is not a valid quantitative index of economic status, there is surely less reason to regard I.Q. as a valid index of cognitive proficiency.

This brings us to an even more fundamental difficulty. Although an index of cognitive proficiency would be considerably more useful than an index of rank order on a battery of tests of unspecified content, it would still not tell us what we really would like to know. Cognitive proficiency is a collection of developing skills in which rates and modes of development are influenced by many genetic and environmental factors. Comparisons between suitable assessments of cognitive proficiency of the same child at different ages tell us something about his intellectual development. Let us suppose that genetic factors determine the modes and rates of development of specific cognitive proficiencies under varying environmental conditions in the way that genetic factors specify reaction norms for physiological characters. Given this model of genetic determination, it is obvious that comparisons between attained levels of cognitive proficiency of different children at the same chronological age cannot tell us much about genetic differences; the relevant information simply is not there.

But the above model of genetic determination is undoubtedly far too simple. Behavioral traits are much more varied in their expression than physiological traits. Intelligence is adaptive, and its manifestations are strongly influenced by cultural and familial environment. Thus cognitive development and proficiency can be meaningfully assessed only within a specific cultural framework.

I.Q. correlation data. The preceding considerations engender little confidence in the potential scientific utility of "psychology's most telling accomplishment." Suppose, however, we follow the example set by Professors Jensen and Herrnstein and ignore the shortcomings of the I.Q. test as a scientific measurement. Suppose, in other words, we agree to regard I.Q. test scores as valid measurements of a biological character like height or weight. What could we then hope to learn from published studies of I.Q. correlation? This question raises two separate issues. The first concerns the applicability of conventional heritability analysis to measurements of a phenotypically plastic trait in natural human populations. The second concerns the quality of the data to be analyzed. If valid scientific conclusions are to be drawn from a given body of data, these data must meet certain widely recognized standards of accuracy, objectivity, freedom from systematic error, and so on. In addition, the theory used to analyze the data must in fact be applicable.

Professors Jensen and Herrnstein lay great stress on the *quantity* of published I.Q. correlation data; they seem less concerned about its *quality*. Recently, however, other scientists have scrutinized the primary sources with a more critical eye. The most penetrating and comprehensive critique is that of Leon Kamin, whose findings are summarized in a recent book.(1) Both Jensen and Herrnstein relied heavily on the work of the late Sir Cyril Burt, whose famous study of identical twins raised together and apart provided the single most impressive piece of evidence in favor of high I.Q. heritability. Kamin has pointed out that, on the basis of internal evidence, Burt's published findings cannot be genuine. For example, Burt reported three measurements of the I.Q. correlation between identical twins reared separately. The first measurement, published in 1955, was based on twenty-one twin pairs; the second, published in 1958, on "over thirty pairs"; and the third, published in 1966, on fifty-three pairs. Yet on each occasion the reported correlation had the same value, .771! Kamin found a number of similar coincidences in Burt's reported data, as well as serious internal contradictions and discrepancies. Jensen has reported that Burt's original data files have been destroyed, so it is not now possible to ascertain the sources of these inconsistencies, contradictions, and numerical coincidences. In these circumstances it is difficult to dissent from Kamin's conclusion that "the numbers left behind by Professor Burt are simply not worthy of our current scientific attention."

The remaining I.Q. data are highly variable in quality. Kamin has made a careful analysis of the most important studies and has concluded that the inferences drawn from them by Jensen and other hereditarians are unwarranted. For example, estimates of I.Q. heritability have been based on comparisons between identical twins reared together and same-sex fraternal twins reared together. Jensen reviewed "all the major twin studies using intelligence tests" and found heritability estimates ranging from forty-seven to ninety-one percent. Kamin notes, however, that Jensen omitted a number of studies that "used sample sizes comparable to the included studies" and seemed equally sound methodologically. When some of these studies are included, the estimates of I.Q. heritability range from an impossible -7% to an impossible +153%. Kamin concludes that the evidence on which the hereditarians have based their case is insubstantial and unconvincing.

Suppose, however, that there did exist scientifically respectable measurements of a genuine biological trait related to intelligence. Could the techniques of conventional heritability analysis be applied to such data? I think not. The concept of heritability does not have an operational meaning for every biological trait, but only for traits that satisfy certain mathematical requirements. I have recently analyzed these requirements in some detail.(2) The main issue is whether the effects of environmental and genetic variation within a population can be disentangled by means of an appropriate mathematical analysis. It turns out that the answer depends both on the nature of the trait and on the kinds of samples that can be drawn from the population.

Most human behavioral traits, including those that contribute to intelligent behavior, may be characterized as *phenotypically plastic*. That is, the development of these traits is strongly influenced by familial, cultural, and other environmental conditions. Language skills are perhaps the most obvious example of phenotypic plasticity. Any normal child can learn to speak any human language, but the language or languages that a child actually learns to speak are determined by experience. It is characteristic of phenotypically plastic traits that the phenotypic effects of a given phenotypic or environmental variation are highly variable within the population. That is, a given environmental change produces different phenotypic changes in individuals who have different genotypes or different environmental histories. Similarly, a given phenotypic difference, as between two fraternal twins for example, gives rise to different phenotypic differences under different environmental conditions. For phenotypically plastic traits, the phenotypic variation in a population cannot be expressed as the sum of a purely genetic and a purely environmental component; a substantial part of the variation must be attributed to the joint effects of genetic and environmental variation.

Nevertheless, heritability can still be defined as the purely genetic fraction of the phenotypic variation. The question now arises: Can this purely genetic contribution be estimated from appropriate measurements of correlations between genetically related persons? The answer turns out to be yes—*provided that, in the sample population for which the measurements are carried out, genotype and environment are statistically uncorrelated.* In experimental animal populations, this condition may be ensured by placing newborn animals in random environments. In natural human populations, on the other hand, it is probably impossible to eliminate genotype-environment correlation. Genotype-environment correlation is always present when children are reared by biologically related persons. For adopted children, the effects are smaller but not entirely absent. Thus it would seem to be a hopeless task to determine the heritability of a phenotypically plastic trait in a natural human population. The data theoretically available in these circumstances are simply not adequate to the task.

Ethical status of studies of I.Q. heritability. Scientific research in the United States as elsewhere is largely supported, directly or indirectly, by public funds. Continued support therefore depends—at least in democratic societies—on popular acceptance of the view that scientific research is in the public interest. And this view will prevail in the long run only if it has a substantial basis in fact. Private foundations are no less sensitive to this requirement than public agencies, because the tax laws under which they operate are explicitly intended to serve the public interest. But how are we to know what kinds of research serve the public interest?

I think it is significant that until recently this question has not been widely discussed in the United States. Nearly everyone agreed that research in the natural and social sciences was desirable—that public benefits are bound to flow from high-quality research, and that the public should therefore support as much of it as it can afford to do. There was also a consensus that authority and responsibility for ensuring that sponsored research was of the highest possible quality, should be decentralized as much as possible so as to preserve for individual scientists and groups of scientists the greatest possible degree of autonomy in their professional activities.

These assumptions are now being questioned in many quarters. Members of Congress have publicly attacked the National Sciences Foundation for funding projects that these Congressmen consider objectionable or unworthy. The Nixon Administration overruled the strong and nearly unanimous objections of the medical-science-research establishment to institute politically motivated and scientifically damaging changes in the allocation of medical-research funds. And from the opposite end of the political spectrum have come demands for the curtailment of research into the genetic basis of I.Q. differences between individuals and groups, on the ground that such research does not serve the public interest.

Now, it may be argued that Mr. Nixon's actions have had a profound and damaging effect on medical research (as in other areas of national life), whereas research into the genetic basis of individual and group differences in I.Q. has not been noticeably impeded by its critics. Yet both Mr. Nixon and those who would proscribe certain kinds of research and teaching as being socially harmful are seeking—admittedly for different motives—to politicize science, to subordinate its traditional values—the values of Galileo, Newton, and Einstein—to some set of political or ethical values. It is my conviction that all efforts to achieve this subordination of science to politics, however nobly motivated, are misguided. Political and ethical considerations are relevant to the *use* of scientific knowledge—not to its *pursuit*.

Let us consider the application of the precept to the I.Q. controversy. Jensen, Eysenck, Herrnstein, and Shockley perceive themselves—and are perceived by a considerable segment of the educated public—as embattled scientists who have been fiercely, even viciously, attacked for publishing and publicizing scientific findings that contradict certain articles of liberal political faith. That the attacks have occurred and that they have often been politically motivated cannot be denied. But the political motivations of Jensen, Herrnstein, Eysenck, and other hereditarians are no less obvious. Consider, for example, the following passage from Jensen's paper on I.Q. and scholastic achievement:

> Since much of the current thinking behind civil rights, fair employment, and equality of educational opportunity appeals to the fact that there is a

disproportionate representation of different racial groups in the various levels of educational, occupational and socio-economic hierarchy, we are forced to examine all the possible reasons for the inequality among racial groups in the attainments and rewards generally valued by all groups within our society. To what extent can such an equality be attributed to unfairness in society's multiple selection processes? ... And to what extent are these inequalities attributable to really relevant selection criteria which apply equally to all individuals but at the same time select disproportionately between some racial groups because there exist, in fact, real average differences among the groups—differences ... indisputably relevant to educational and occupational performance?

Again, consider Jensen's discussion in the same article of the alleged failure of compensatory education. Compensatory programs have failed, says Jensen, because they were designed in the light of fallacious theory. Now, a layman might suppose that the basic assumption underlying compensatory programs is that every normal child is capable, under appropriate conditions, of acquiring the cognitive skills he or she needs to participate effectively in a modern technological society. Educability in this sense is obviously both a necessary and a sufficient condition for the eventual success of compensatory programs. Professor Jensen, however, as a professional educational psychologist, perceives two much more specialized assumptions underlying compensatory programs: 1. that all normal children "are basically very much alike in their mental development and capabilities," and 2. that "those children of ethnic minorities and the economically poor who achieve 'below average' in school do so mainly because they begin school lacking certain crucial experiences which are prerequisites for school learning...." Jensen's eagerness to write off compensatory programs has here led him to caricature his opponents' position in a way that seems more appropriate to a high school debate than to a scholarly article. Of course, as an educational psychologist attacking educability, he is in a delicate position; but I find it hard to believe that he has been forced into that position by disinterested scientific considerations.

In fairness, I think it must be granted that both sets of participants in the I.Q. controversy have been more or less strongly motivated by political and ethical considerations. The hereditarians tend to believe that it is futile as well as immoral to tamper with the Natural Order laid down in the chromosomes; their opponents believe in educability and in greater equality. It is not surprising that members of the two groups disagree as to the ethical status of research aimed at elucidating the biological sources of individual and group differences. Nor do I know of any *a priori* ethical principle that would resolve this conflict. On the other hand, I am moderately optimistic that the area of conflict will steadily diminish as an understanding of the underlying scientific issues becomes more widespread. In this connection the I.Q. debate has rendered and continues to render an important public service. It has forced scientists on both sides of the issue to take a hard look at certain substantive and methodological assumptions that they have accepted uncritically all their working lives. It is precisely because these assumptions have important social and political implications that their scientific validity is now being questioned. A colleague of mine, who happens to be a psychologist, once complained that the critics of heritability analysis as applied to I.Q. scores did not object when the same techniques were applied to such behavioral traits as extraversion or aggression. This showed, he said, that our criticisms were politically motivated. He was right, of course. What he failed to notice was that the validity of a scientific criticism is unaffected by the critic's motivation.

To sum up, I would insist on the distinction between the *pursuit* of basic scientific knowledge, and its *use*. The use of basic scientific knowledge often raises serious ethical questions, but I regard the pursuit of pure knowledge as good in itself—even when it teaches us things we would rather not know. Yet scientists who address themselves to socially or politically sensitive issues have a special responsibility. If I, as an astrophysicist, publish a mistaken estimate of the abundance of helium in a stellar atmosphere, I injure only my own reputation; the star will survive. But the social scientists' mistakes may have more serious consequences; they are confronted with the problem of assessing the validity of research in psychology and the social sciences.

THE PROBLEM OF VALIDITY

In pure mathematics, the problem of validity scarcely arises. Occasionally, a faulty deduction in an especially complex proof may go undetected for a period of months or even years, but in principle it is always possible to check the validity of a proposed proof. The well-known controversy between formalists and intuitionists concerns not the validity of specific theorems but the admissibility of certain logical axioms.

In the physical sciences the situation is somewhat more complex. Physics is based on a small, slowly evolving set of mathematical laws. The predictions of physical theories are logical inferences from these laws. Hence the mathematical correctness of a prediction can be verified in precisely the same way and by the same methods as the validity of a theorem in pure mathematics. But physical laws are required to have another kind of validity: their predictions must "agree" with experiment or observation. It is not easy to specify exactly what is meant by "agreement" in this context. Because all measurements and observations are subject to error, they can never agree precisely with the predictions of a mathematical theory. Moreover, no existing theory applies to all physical phenomena under all circumstances. Newton's theory of gravitation, for example, applies with an enormously high degree of precision to a vast range of gravitational phenomena, from the motions of the planets and their satellites to the motions of galaxies in giant clusters, yet it fails to describe the expansion of the universe as a whole or the dynamics of highly collapsed stellar masses like neutron stars and black holes. Moreover, there are many phenomena to which existing physical laws are thought to apply but which nevertheless are not yet fully understood. For example, although the structure of atoms and molecules is thought to be governed by known laws, only the simplest atoms and molecules have received a theoretical description of comparable accuracy to the experimental measurements. Again, although the laws of classical physics are thought to apply to the primordial solar system, we have as yet no satisfactory theory for the origin of the planets. Yet both theories and experimental or observational findings must meet exacting standards of rigor and objectivity before they are accepted by the scientific community. Strict adherence to such rigorous standards has made possible the rapid cumulative growth of the physical sciences during the last three centuries.

The biological sciences differ methodologically from the physical sciences chiefly through the greatly diminished role of mathematics. In this respect biology is not unlike the physics of highly complex systems like planetary atmospheres or large molecules. Yet the biologist is as concerned as the physicist with experimental rigor and with causal explanations based on well-established physical principles. With the burgeoning of molecular biology, the natural sciences have become an integrated whole composed of diverse but interpenetrating parts and governed by a single set of methodological standards.

When we come to the behavioral sciences, the outlook changes. Although no one will dispute that human behavior, as a biological activity, forms part of the subject matter of the biological sciences, not all behavioral scientists are prepared to accept the methodological standards and philosophic outlook that characterize the mature natural sciences. Behaviorists, for example, insist that only *overt* behavior can be studied scientifically and that statements about unseen mental states and processes are scientifically worthless. This doctrine is reminiscent of the objection that some people raised to the molecular theory of gases around the turn of the century. These people argued that the molecular theory was inadmissible because no one had ever seen a molecule. Molecules are not, indeed, objects of perception, but scientific constructs, which need not relate *directly* to experience.

Scientific constructs must, however, meet exacting standards of logical rigor and scientific relevance. Unfortunately, not all psychological constructs satisfy such requirements. The psychometric definition of intelligence, for example, lacks both logical rigor and biological relevance. I do not by this remark mean to impugn the clinical value of psychological testing, still less to suggest that psychological tests lack "validity" or "reliability" in the technical statistical senses of these terms. What I do assert is that the psychometric definition of intelligence does not yield the sort of construct that can figure in a causal or predictive description of behavior.

A biological approach to intelligence would begin by recognizing its essentially adaptive and plastic character. Intelligent behavior is manifested not so much in the way an animal copes with a fixed set of environmental challenges but by the ability to devise effective responses to new and unforeseen environmental challenges and to make creative use of relevant past experience. Thus intelligence is not a property of behavior as such but of behavior viewed in the light of previous experience. It follows from these elementary considerations that no meaningful comparison with regard to intelligence can be made between persons reared in different cultures or subcultures. The very notion of a culture-free test of intelligence reveals a fundamental—and unfortunately very widespread—misunderstanding of the biological concept of intelligence. There are culture-free tests and there are tests of intelligence, but the notion of a culture-free test of intelligence is a contradiction in terms.

Even comparisons between persons reared under closely similar conditions are fraught with difficulty. Peter and Paul are identical twins. At age twelve, Peter breaks his leg trying to set a new downhill record on Mt. Snow. While recuperating he comes across a book of mathematical puzzles and is inspired to spend most of his waking hours during the next six years developing his mathematical proficiency. Meanwhile, back on Mt. Snow, his equally compulsive brother Paul pursues a skiing career with equal single-mindedness. How are we to compare the adequacy of Paul's response to environmental challenge with Peter's? And how are we to go about disentangling the effects of compulsiveness from those of intelligence? Questions like these show that the common-sense notion of intelligence refers to a complex and ill-defined aggregate of biological characters. As a science matures, such fuzzy, ill-defined notions tend to disappear; like chickenpox, they are diseases of infancy.

So far I have stressed the continuity that must obtain between biology and the behavioral sciences if the label "science" is to have more than honorific significance. I have stressed both methodological and substantive aspects of this continuity. While some psychologists have, in my opinion, paid too little attention to the biological foundations of their subjects, some biologists—and certain psychologists as well—have paid insufficient attention to the most distinctive feature of human behavioral traits—their developmental plasticity. The distinction between phenotypically stable and phenotypically plastic traits is not entirely one of degree. For some physiological and behavioral traits, deviations from an optimum physiological structure or behavior pattern may substantially reduce fitness. Natural selection then favors developmental homeostasis or canalization. Such traits are comparatively insensitive to moderate environmental variations. For other traits, however, natural selection favors plasticity over stability. This is particularly true of the behavioral traits that concern psychologists. A capacity for language is part of the genetic heritage of every normal human being. But linguistic competence is so protean in its manifestations that no one has yet succeeded in characterizing its genetic component. This example suggests that the development of plastic behavioral traits may differ *qualitatively* from that of relatively stable physiological traits. For this reason, biologists and psychologists should, I think, exercise great caution in applying generalizations that have been verified for the first kind of trait to the second.

In these circumstances, we have no reason to expect that an analytic technique like heritability analysis, developed and tested in a limited biological context and based on highly restrictive, if implicit, assumptions concerning the developmental process, can meaningfully be applied in the qualitatively different context of plastic behavioral traits. Moreover, lacking an understanding of how such traits develop, one must be exceedingly cautious in making predictive generalizations. Induction is not enough, since it affords no basis for prediction under circumstances that differ significantly from those already experienced. For example, generalizations concerning the existence of qualitatively distinct cognitive styles have no predictive force. I have in mind Professor Jensen's suggestion that children who score below a certain level on I.Q. tests lack the capacity to develop problem-solving and abstract-reasoning abilities. Such suggestions should not be treated as scientific hypotheses. So far as I can see, they serve no positive function whatever.

To sum up, the scientific study of behavioral traits does call for distinctive techniques and for new scientific insights. Psychology cannot be reduced to biology any more than biology can be reduced to physics. Yet physics, biology, and psychology are linked by powerful methodological and substantive bonds. Psychological theories must make sense biologically, just as biological theories must make sense physically. And assertions about human behavioral traits must be backed up by theoretical considerations and experimental data that meet the standards of rigor that prevail in the natural sciences. Otherwise there is no reason why the educated public should pay serious attention to them.

THE BEHAVIORAL SCIENCES AND SOCIAL RESPONSIBILITY

I have argued that the ethical status of scientific research cannot profitably be discussed until its scientific status has been clearly defined. The only moral issue presented by bad research is whether it should be actively supported—and that is an issue on which there is likely to be little disagreement. What about scientifically sound research which in its application or implications may be socially harmful? This is a much thornier question, for which, I suspect, there is no generally valid answer. To ban all scientific research possessing potentially harmful social consequences would be to ban scientific research altogether. On the other hand, most people would agree that some specific kinds of applied research do not deserve public support. Between these two extremes there is an area in which scientific and sociopolitical considerations may come into conflict. Here each case must be considered separately in the context of broader scientific and social philosophies. The scientific community can contribute most constructively to such discussion by helping laymen to a better understanding of the purely scientific quality and importance of the research in question.

Among physical and biological scientists there seems to prevail a broad consensus on questions of scientific quality and importance. Of course, the views of individual scientists and groups of specialists are inevitably biased in predictable ways. Yet in spite of conflicting special interests—in spite of the pluralistic character of scientific research—physical and biological scientists seem to share basic scientific values and priorities.

I detect no such consensus among behavioral scientists. Between behavioral scientists working in closely related fields, there seems to be little communication or even sympathy. If natural scientists make up a community, behavioral scientists seem to inhabit a collection of walled villages. Psychology, which has—or ought to have—closer ties with the natural sciences than the other behavioral sciences, is the most fragmented. There is scarcely more interaction among Skinnerians, Freudians, and Piagetians than among Buddhists, Christians, and Marxists.

The radical disunity of the behavioral sciences is a reflection of their comparative youth and of the enormous complexity and diversity of their subject matter. Psychology is inherently a more difficult subject than astrophysics because human behavior is incomparably more complex than the behavior of self-gravitating gas masses. To recognize that the behavioral sciences are immature and fragmented is not, however, to deny their importance or to belittle their accomplishments. It is merely to recognize that these accomplishments do not yet afford a reliable basis for *prediction*.

In the physical sciences, prediction first became possible with the formulation of general dynamic laws. In organismic biology, the theory of organic evolution provides a conceptual framework that unites all aspects of the subject and makes possible certain qualitative and semi-quantitative predictions. The behavioral sciences are still largely descriptive. Theories abound, but there is as yet no single all-embracing Theory comparable to the great unifying principles of physics and biology, and without such a theory no genuine prediction is possible.

This point seems not to have been fully grasped by many behavioral scientists. Skinner, for example, has argued that theory is actually an encumbrance to the behavioral scientist, who ought to approach data with an open mind. If the scientist has chosen the right variables to measure, his measurements will exhibit lawlike regularities that will enable him to predict the outcomes of future experiments. The flaw in this Baconian fairy tale is that predictions based on such empirical regularities nearly always fail unless the circumstances of the original experiment are precisely reproduced. This is true, for example, of the "laws" that Skinner and his followers induced from their experiments with rats and pigeons: these "laws" cannot be reconciled with the learning behavior of newborn human infants. (3)

Empirical rules are not useless, of course. They may help a clinical psychologist to spot incipient paranoia, or an industrial psychologist to spot likely candidates for the job of office manager. In this sense empirical rules may be said to be predictive. They are predictive in the same way that weather forecasts based on the assumption of persistence are predictive—and they have the same limitations. They tell us that closely similar antecedents are *usually* followed by similar consequences, or that if n-1 elements of a complex picture are present, then the nth element is also likely to be present. There is no gainsaying the usefulness of such rules of thumb; but they must not be confused with predictive laws like those on which the natural sciences have been built.

A truly predictive law always goes far beyond the phenomena it was initially devised to describe. Thus Newton's law of gravitation not only made possible a more accurate description of planetary motions than Kepler's laws (which, despite their mathematical form, are merely empirical rules) but also explained a large number of qualitatively different phenomena, such as the tides.

It cannot be emphasized too strongly that the natural sciences, and especially the physical sciences, owe their predictive success neither to their adherence to a specific methodology nor to their use of mathematics but to the discovery and exploitation of fundamental mathematical laws underlying the phenomena they seek to describe. No amount of experimental rigor (even if it could be attained in the behavioral sciences) or statistical anlaysis can make up for the lack of an adequate theoretical foundation. A leading theoretical physicist, Richard Feynman, made this point in a commencement address at Caltech in June 1974:

> I think the educational and psychological studies I mentioned are examples of what I would like to call Cargo Cult Science. In the South Seas there is a Cargo Cult people. During the war they saw airplanes land with lots of good materials, and they want the same thing to happen now. So they've arranged to make things like runways, to put fires along the sides of the runways, to make a wooden hut for a man to sit in, with two wooden pieces on his head like headphones and bars of bamboo sticking out like antennas—he's the controller—and they wait for the airplanes to land. They're doing everything right. The form is perfect. It looks exactly the way it looked before. But it doesn't work. No airplanes land. So I call these things Cargo Cult Science, because they follow all the apparent precepts and forms of scientific investigation, but they're missing something essential, because the planes don't land.

One form of Cargo Cult Science is so prevalent among behavioral and other social scientists, and so widely respected, that it deserves brief discussion. I refer to the use of certain statistical techniques that analyze the variation within a given population of a dependent variable (for example, subject's income or I.Q.) into components "caused by" or "attributable to" variations in certain independent variables (for example, father's income or mother's educational level). The data used in these analyses are statistical correlations among the variables: for example, income or I.Q. correlations between genetically related persons. Although the various techniques currently in use differ in detail, they all depend on certain key assumptions: 1. that factors possessing variation affect the variable being studied can be identified; 2. that they can be quantified and measured; and 3. that all relevant causal relations are linear.

Identifying the relevant causal factors. At first sight this seems straightforward enough: Simply list all the factors you think might be relevant. If you can buy enough data and computer time, the computer will tell you which ones actually are important. Or will it? Suppose you are studying the effects of education on income (as Christopher Jencks and his collaborators recently did), and your computer tells you that variations in schooling have little effect on variations in income. Are you then justified in concluding, as Jencks and his collaborators did, that quality of schooling has little effect on adult income? A second example makes it clear why this conclusion cannot be drawn. Suppose an eighteenth-century statistician has set out to investigate the causes of postpartum maternal mortality. He would almost certainly have found no evidence that the quality of medical services was implicated, because the specific medical procedure responsible for childbed fever was not isolated until the nineteenth-century. This example illustrates a general rule: From a negative statistical finding, one can infer only that one has failed to identify the specific causal factors that are relevant to the inquiry. Thus the findings of Jencks and his collaborators do not imply that variations in schooling have little effect on variations in income, but only that—as Skinner might point out—the study did not focus attention on the right independent variables. But in the absence of an underlying theory that relates education to income, one has no guide to the choice of independent variables.

Quantifying and measuring the variables. Most of the variables behavioral and other social scientists are interested in are not measurable in the same sense as physical quantities like height, weight, and temperature. It seems not to be widely understood by social scientists that measurement involves more than a systematic assignment of numbers to experimental outcomes. It is easy enough to invent a consistent scheme for converting appropriate information about income, job education or style of dress, into a numerical index of socioeconomic status. The trouble is that different people will devise different indices. One person may attach a lot of weight to formal education, another to income, a third to the social graces. (Precisely the same considerations apply to the "measurement of intelligence.") Many social scientists believe that measurement confers objectivity. What they sometimes fail to recognize is that the quantification of inherently non-quantitative variables introduces a complex and unanalyzed assortment of value judgments and prejudices.

The assumption of linearity. Even when the relevant variables are well-defined and measurable, as in econometrics, the third major assumption underlying analyses of variance is rarely valid. Generally speaking, the assumption of linearity is reasonable only for variations so tiny as to be of no practical interest. When variations are large enough to be interesting, the assumption of linearity nearly always breaks down. Terms like "feedback," "regulation," and "saturation" all refer to nonlinear causal relations. In real life such nonlinear effects are usually important. This explains why the statistical techniques under discussion do not form part of the repertoire of physical scientists, who also deal with populations and who also would like to have foolproof techniques for disentangling causal relations.

How, then, is the social scientist to go about discovering causal relations? I suggest that the natural sciences do provide a useful model—provided that one focuses attention on what natural scientists do rather than on what conventional wisdom says they ought to do. The solution to a problem in the natural sciences usually proceeds in four stages. 1. The scientist studies the problem long and hard enough to achieve some fresh insight into it. 2. He uses this insight to formulate a definite hypothesis. 3. The implications and empirical consequences of the hypothesis are then worked out. 4. He devises observations or experiments that could disprove his hypothesis by contradicting one of its implications. This procedure—the hypothetico-deductive method—does not require the hypothesis to have a mathematical form, nor does it presuppose the existence of underlying mathematical laws. The evidence relevant to a hypothesis *may* be quantitative or statistical, but it need not be. What is essential in a scientific hypothesis is that it be significantly simpler than the data it serves to organize, and that it make unambiguous predictions that go beyond these data. Of course, a great deal of research in the behavioral sciences meets these requirements. The crucial step, omitted or slighted in the studies I have singled out for criticism, is the first one: the search for genuine insight. Because there exists no prescription for acquiring insight, there is no method or technique that will enable either a natural or a social scientist to solve genuine problems. "Research"—the acquisition and organization of relevant data—is of course essential, but it is not enough. If there is a lesson that social scientists can learn from natural scientists, this is it.

To sum up, I believe that there do exist objective grounds for assessing the validity and importance of research in the behavioral and other social sciences. Although my experience and understanding in this area are limited, I am acquainted with a number of studies that seem to me to be of the highest scientific quality. Yet, because the behavioral sciences do not yet possess the same kind of secure theoretical basis as do physics and biology, their predictive scope is limited. The insights they furnish cannot safely be extrapolated to situations very different from those in which they were formulated. For this reason the behavioral sciences are not yet in a position to make strong assertions about the *limitations* of individual human behavior or of human institutions. Statements like "Society must accept man's innate aggressive tendencies as a biological given" or "A child with an I.Q. of eighty-five can never be taught to reason abstractly" are scientifically unjustified extrapolations. To recognize such limitations is not to depreciate the achievements of the behavioral sciences; not to recognize them is bound to have harmful social consequences and will end by bringing the social sciences into disrepute.

REFERENCES

(1) L.J. Kamin, *The Science and Politics of I.Q.* John Wiley & Sons, New York, 1974.
(2) D. Layzer, "Heritability Analyses of I.Q. Scores: Science or Numerology?" *Science*, 183, pp. 1259-66 (1974).
(3) T.G.R. Bower, *Development in Infancy.* W.H. Freeman, San Francisco, 1974.

SOCIAL MARGINALIZATION: TOWARD A GENERAL THEORY OF INEQUALITY

Harland Padfield

Social science and policy in the last two decades has had two compelling concerns—on the one hand how to explain and reduce non-assimilation and inequality of people in the national mainstream system of benefits, and on the other hand, how to increase our understanding and discrimination of human abilities to perform in the national production system. It is the central tenet of this paper that in both of these traditions we are dealing with one phenomenon and further, that scientific performance in both pursuits contributes generally to the social and economic forces at the root of inequality.

SYSTEMATIC BIASES FROM THE KNOWLEDGE MARKET

Although few scientists would deny that scientific work responds to incentives, many would argue that incentives vary with the individual, that biases are variable and counter balancing and that gaps in knowledge are random. But

> ... if scientific knowledge and professional work is increasingly important for defining and solving public problems, it is also true that highly organized centers of power substantially influence what is defined as problematic, that for which knowledge is sought, and acceptable solutions. Although the public has become more and more dependent on science, science has become increasingly dependent on the resources of a few.
>
> ... knowledge that develops and the problems that are investigated are those of direct concern to the centers of power. One of the most powerful

institutions that has influenced scientific work and its content has been the modern industrial corporation.

...corporations invest in research primarily for production development. Consequently, conditions such as housing, pollution, and general environmental deterioration receive little attention. Organizational goals become the primary determinant of scientific work rather than public need.(1)

The concentration of economic power affecting the development of knowledge could not occur without cooperative development of centralized regulatory power in the public administrative bureaucracy. In short, a relatively few people who are not scientists are in a position to make enormous resource allocations to the conduct of scientific inquiry. One inevitable result has been more stringent and explicit cost/benefit constraints and the quantum increase of single purpose mission-oriented research as opposed to curiosity satisfying, diffuse and comprehensive purpose research. These dynamics derive from the rational decision rules and political imperatives by which industrial corporations and government bureaucracies exist.

In terms of effects on the behavioral or social sciences, systematic pressures dictate concerns about humans in certain capacities and not others. For the institutions in the business of using human resources, human characteristics which provide clues to how people will perform in the production process are of direct concern. For those institutions in the business of socializing or investing in human resources, being severely constrained by tax payers' cost accounting, the efficient handling of people in terms of the institution's production goals also becomes inevitably rational. The consequence of these dynamics is an enormous convergence of interest in how to capture human resource investments already made by other institutional situations—those made in the natural course of individual growth and development. Thus incentives are created to develop and continuously refine the science of identifying, selecting, and enhancing human resources appropriately adjusted and cognitively consonant with the national mainstream system, while disincentives are created to develop a system to rehabilitate and retrain human resources, including linguistic and racial minorities, who are maladjusted and cognitively dissonant with the mainstream system. *Processing* rather than creating or recycling human resources thus becomes the primary mission of the educational establishment and the primary purpose to which social science research and knowledge are put.

SYSTEMATIC BIASES FROM THE PROFESSIONAL SYSTEMS

Response to pressures created by national economic and political institutions are not the only biases in the social sciences affecting cultural minorities. Charles C. Gillispie, a Princeton University history of science professor, comments in a recent review of Robert K. Merton's *The Sociology of Science*(2) that the main thrust of Merton's analyses is clearly and convincingly that scientific

... behavior occurs in the service to social norms; that norms arise in the life of real communities governing the conduct of their members; that the phrase "scientific community" is, therefore, no mere manner of speaking about some shared pleasure in the study of nature but refers to an effective social entity; and that, within its membership, which is bounded professionally and not geographically, two main sets of norms constrain behavior and do so in ways that conflict, the one enjoining selflessness in the advancement of knowledge, and the other ambition for professional reputation, which in science accrues from originality in discovery and from that alone. The analysis exhibits the scientific community to be one wherein the dynamics derive from the competition for honor even as the dynamics of the classical economic community do from the competition for profit. . . .

If selfish desire for recognition and honor are indeed the basis of incentives operating in the scientific community then it is also reational and inevitable that the knowledge generated by social science will in the main address the issues of central interest to the discipline professionals. This knowledge will tend to elaborate and extend conventional theoretical constructs rather than challenge them, and where work is applied, it will tend to define problems and dictate solutions that are politically and economically acceptable. It is not this generally recognized fact which I wish to dwell on, but how the peculiar biases of the disciplines have converged to create systematic distortions in our constructs of reality, particularly as they apply to the experiences of subordinant cultural minorities. I write mainly of anthropology/sociology, psychology, and economics.

Without recapitulating or abstracting the numerous critiques of the "culture of poverty" tradition in anthropology and the lower class culture theory in sociology, I will simply summarize general deficiencies and distortions in these bodies of literature.(3)

Probably the most serious general bias is the tendency to operate with the priori assumption of culture as a system for determining behavior rather than as a system for adapting to effective economic and social environments. Coupled with this is the anthropologists' penchant for over emphasizing the unique characteristics and internal dynamics of cultural groupings as opposed to the external dynamics and general similarities operating with respect to the society at large thus reinforcing apriori assumptions of cultural causation. Oscar Lewis demonstrates this repeatedly despite his theoretical statements as to the generalizability of his data. To Lewis, however, goes the credit for establishing a major focus of interest among anthropologists and for developing the theory of cultural dynamics of the situation of poverty.

Anthropological biases have been reinforced generally by sociological studies of the lower class in the theoretical context of urbanization and social deviancy. Again the internal dynamics are emphasized by explicit theoretical frameworks which require data on the behavior of the lower class subjects—many times from the regulatory agencies and institutions which oversee them.(4)

The biases of the social sciences derive in large part from the way the disciplines have partitioned social reality—i.e., philosophically defined their domains. The central issue is how their partial constructs of reality affect their respective encounters with economically and politically subordinate minorities. Furthermore, it can be argued that their biases are mutually reinforcing rather than counter balancing because each discipline's constructs inherently obviates environmental variation. While anthropology focuses on cultural uniqueness, it tends to treat the physical and social environments of cultures as given. And while psychology focuses on individual uniqueness, it tends to be preoccupied with primary institutions and also tends to accept larger social systems such as political and criminal justice institutions, public education and welfare, private medicine, organized labor and industry as natural. Thus both anthropology and psychology tend to ignore the unique environments created by macro institutions vis-à-vis race and class minorities.

Perhaps the most important foundation for bias in the social science disciplines is the fact that each professional community has come to be identified with what each likes to regard as a separate, analytical reality. This is despite the fact that methodologically all social sciences depend upon natural, complex behavioral systems, none of which can be thoroughly understood without all disciplines including economics. The compound result of this philosophical discontinuity has been that each discipline rather than being limited to its proper analytical domain has instead been allowed to assume professional proprietary rights over a *natural* system and in the process come to be allowed to speak authoritatively concerning the complex whole with which it is dealing. Thus enormous blind spots develop which are inevitably filled by conventional (usually culturally conditioned) assumptions the scientist has about "human nature."

To re-emphasize, the result is compound bias—i.e., not only is there disciplinary bias with respect to assumptions about how these systems operate as a whole, but there is culturally imbedded bias with respect to assumptions about how they operate in different social settings. Thus, economics has generated partial explanations regarding the behavior of firms, but leaving the behavior of "economic institutions" to the interpretation of economists implies, to the public's detriment, that other institutions such as the family, the school, and even the church do not have major economic functions.

Psychologists' analytical domains are individual intellectual, cognitive, and affective systems and micro-interaction systems as operant situations; whereas their proprietary domain is the U.S. middle class person, either in natural or experimental settings. But whether humanist or behaviorist, psychologists as theoreticians of individual behavior (with the help of anthropology and sociology) complement the compound bias of economics—humanist by tending to view the family and individual as identity systems as opposed to rational systems with objective goals; both humanists and behaviorists by tending to generalize American middle class models cross-culturally. The result is that when the principles of behavior fail to work cross-ethnically and culturally, the tendency is to incorporate constructs from anthropology and sociology and attribute maladaptive behavior to cultural conditioning and social deviancy.

The upshot is that scientific understanding of lower class behavior and public policy toward it has contended with an enormous bias generated by the social science disciplines from the time of their very formation. Without taking or joining issue with others who have written on this(5), the consequences I wish to summarize at this point are philosophical rather than social. *The complementary sensitivities and insensitivities of the social science disciplines have persistently converged to over emphasize variability among ethnic, class, family, and personality systems, while preconceiving social-environmental uniformity.* This means that in the conduct of scientific consultation to social policy, the blind are leading the blind concerning the differential operation of our basic economic and public institutions in the midst of intense scrutiny of their workers, students, cases, clients, patients, inmates, and general outcasts. Thus, it would come as no shock to foreign scientific observers of the American system that the so called "enlightened" social policies of the sixties failed, and that we are now engaged in a compelling re-examination of discredited or partially discredited social scientific tenets.

It is my contention that the partial failures of the social programs of the sixties occurred not because of illusions concerning the magnitude and plasticity of differences among races and classes of people, but because of fundamental illusions concerning the *co*-operation of basic economic and public institutions with these differences in a dynamic and continually evolving social system.

MARGINALIZATION VS. ASSIMILATION

In Milton Gordon's classic study *Assimilation in American Life* (1964), he defines the "ideal type" or complete state of *assimilation* by describing a hypothetical example of a host country with the fictitious name of Sylvania where race, religion, and previous national extraction are the same and cultural behavior is relatively uniform *"except for social divisions"* and where the groups and institutions are differentiated *"only on a social class basis."* He introduces another hypothetical group called the Mundovians into this country by immigration who by the second generation are no longer distinguishable racially, culturally, or structurally from the rest of the Sylvanian population. In Gordon's words, becoming assimilated in Sylvanian society means the Mundovians have: Changed their cultural patterns to those of the Sylvanians; entered fully into their societal network; intermarried and interbred fully with them; developed a Sylvanian ethnicity; no longer encounter discrimination or prejudice; and are not in political conflict with them.(6)

The need to explain cultural processes tending toward uniformity was of central concern to social anthropologists and cultural sociologists in the fifties. Since then, however, their attention has been directed to cultural processes tending toward diversity and social variability because events have made it all too apparent that quite the opposite of Gordon's assimilation model has occurred throughout culture contact. Or putting it another way, the processes set in motion when two cultural systems come into contact—especially dominant/subordinant—are exceedingly complex having no end point only stages, phases, and aspects with varying degrees of equilibrium and closedness in time and space.

If I were to suggest an oversimplified antithesis in Gordon's model it might go something like this: Several centuries and 10 to 20 generations since the Indians, Africans, and Mexicans encountered the "Sylvanians," a preponderant majority have neither entered the Sylvanian mainstream cultural system nor have they been able to maintain their original cultural system. Thus, the overly simplistic assimilation model A + B = A gave way to cultural pluralistic models A + B = AB, C + AB = ABC, etc. as exemplified in Moynihan and Glazer's book, *Beyond the Melting Pot*.(7) However in the face of disquieting persistence of intergenerational poverty among ethnic and class minorities, the cultural pluralistic model appears to distort reality as well. Therefore, instead of attempting to develop models of assimilation or cultural pluralism, we should be attempting to develop models as illustrated by: A + B = Ab, C + Ab = Abc, etc. which suggest movement from a geographically separate and culturally distinctive position to restricted or limited participation in the politically dominant society.

Perhaps several generations past, prior to continuous, broad scale socialization in public institutions, this condition or process could be referred to as cultural pluralism. But under increasingly prevailing conditions analyzed below as saturation human capital investment, members of ethnic and class minorities are experiencing debilitating socialization—i.e., much of their cognitive and normative experience within the structure of public institutions is useless or has negative benefit. And as a result of the economic differentiation associated with these "investment" experiences, increasing numbers find themselves in cultural enclaves as adults living out their lives—including the responsibility of having to socialize their children—in neither their native culture, nor mainstream American culture, but in what can more accurately be termed the culture of marginality. Correspondingly, rather than assimilation or pluralism, the processes affecting them are *social marginalization*.(8)

My discussion of these processes begins with an analysis of the operational environments within which adaptive social units (presumed to be rational) must develop viable, institutionalized responses.

THE REGULATION OF EFFECTIVE ECONOMIC ENVIRONMENTS

In a recent demographic study of "Institutions in Modern Society," Octavio Romano concluded that on any given day 40 to 45 percent—almost half—of California's total population was subject to some form of public institutional regulation.(9) The four service systems Romano defines in this category include schools, social welfare, law enforcement, and hospitals. All are involved in socialization, enculturation, or as economists say *human capital investment*. My emphasis is not the pervasiveness of this function but its magnitude. Not only do virtually all citizens of modern affluent societies experience publicly administered socialization in the early stages of their lives, they continue to experience it in some form virtually all their lives. Considering public broadcasting, public manpower training, and other forms of occupational training in industry, as a citizenry we are virtually saturated in human capital investment.

This brings me to my major point concerning the inadequacy of human capital, socialization and enculturation theory to explain variability in social and economic welfare—i.e., why heavy investment in public education and other public services in and of itself does not eliminate systematic economic disparity and social inequality. This requires a clear examination of the key institutional systems involved in the investment, processing, and management of human resources. These include public education, the labor market, criminal justice, public health, and social welfare.

Virtually every health, education and manpower program is predicated, in part, on the doctrine of the operation of a free competitive human resource market—as opposed to a highly regulated, discriminatory market—and the naive expectation of system inertia in the face of rapid economic development and social change. Thus repeatedly, educational psychologists and economists view the educational system as a human investment institution. This is fine except that it is also implicitly assumed to be egalitarian in the investment process. Perhaps it *was* egalitarian in terms of its client groups at some point in the historical past when the administration of educational benefits occurred as a natural by-product of the welfare-rationing function of other social and economic institutions. But when the national educational system is deliberately transformed as a matter of public policy into a truly mass institution, the disparate investment function which it was performing *de facto* in the total context of inequality in the U.S. could be maintained or "protected" only with the concurrent development of a human resource *processing* function—i.e., testing, grading, sorting, and allocating human capital by means of rationing economic credentials.

In this conceptual framework the study of educational "problem populations" becomes less interesting than the study of how the public educational system administers this function and how the evolution of this function interacts with changes in other economic institutions. This would more than likely bring the problem of school failures into scientific focus as the study of adaptive behavior in a newly evolved social environment in a basically unchanged social structure.

Similar illusions underlie manpower policies where the overwhelming emphasis has been on altering the behavior and occupational competencies of the unemployed thereby increasing their chances of becoming selected in the labor market. The simplistic notion is that labor markets are keyed primarily to screen the labor pool for the most productive competencies independently of noneconomic factors. The one major federal manpower program which demonstrated conclusively that labor markets don't operate this way was the urban ghetto-oriented NAB/JOBS program launched amid urban unrest in 1968. Direct employment of "unqualified" people in industry was subsidized thus effectively altering the discriminatory function of the labor market. Between 300,000 to a half million ghetto unemployed were exposed to industry with 2/3 becoming converted in the process. Despite its relative success, the program was ended by economic recession and union pressure less than two years after it began.(10)

With few exceptions, manpower research emphasis continues to focus on the behavior of the unemployed as opposed to the more interesting inquiry into the dynamics of job rationing in industrial labor markets and public regulatory agencies. With this kind of theoretical focus, the behavior of the hard core unemployed falls hypothetically into place as resilient adaptation to an economic environment with disincentives to investment as profound as those faced by business in a global depression.(11)

In not only the public but the private sector as well, the public administration of benefits is increasingly decisive in economic success. Thus the dairy industry evolves from a community of producers of milk competing simply for profit on the open market to a complex quasi-public association producing milk for a government subsidized market and lobbying for price supports in an effort to capture the benefit of public policy. Thus a contribution to a political candidate in a key position to control the federal agencies regulating the dairy industry is as much an investment as the building of a new creamery.

Clearly a broader definition of investment in both the business and human capital sense is necessary. The investment concept should include diverting present income to increase skills and capacities to capture *administered* benefits or income. Thus economic disparity and social inequality are a function of institutionalized differential capacities to capture administered benefits.

Thus labor markets must be seen as complex systems that function as importantly in the distribution process as the production process. Rationing mechanisms are operating in which industry and other economically organized groups—especially unions and the professions—continually invest to maintain or increase their own (as opposed to the public's) economic well being. These rationing mechanisms are linked directly with public education.

In the context of the distribution of economic welfare in a technologically dynamic society, *school is primarily a system for the development and administration of differential credentials by means of which economic opportunity and social statuses can be rationed by the labor market.* Economic credentials are administered more or less consistently with the differential preferences for cultural and racial characteristics established by the users and organizers of labor. Among other things, the learning process in school involves the mapping of these preferences and the development of corresponding incentives and disincentives.

FAMILY AND SOCIAL NETWORKS AS ECONOMIC INSTITUTIONS

When economists discuss the woes of a business community, the economic environment in terms of incentives/disincentives is generally recognized as an independent set of variables and the policies and practices of the industry in question as a dependent set. When the health of an economic sector or a major industrial corporation is at stake, restructuring the economic environment is invariably called for. Thus operating within the deductive framework of economic theory, the industrial corporation is assumed to be a rational system with specific goals and a core technology—including formal organization, behavioral codes, and a corporate rationale—for the achievement of these goals. Given the overwhelming importance of production goals and core technologies, new kinds of opportunities and constraints tend to compel decisions opposing changes that will disrupt the internal technology.(12) Whether this technology has been organized around manufacturing for maximum profit, gasoline, automobiles, munitions, opium poppies, USDA subsidized crops, etc. makes little difference.

Drawing general logical inferences from this system of reasoning, it is necessary simply to recognize public policy as the effective environment, the industrial corporation as the adaptive unit and core technology as the adaptive strategy The economists are the consultants to public policy and macro-economic policy changes are simply good behaviorist principles otherwise known as creating a "sound business climate."

In applying this economic-behavioral framework to the family in a logically consistent manner, we need simply to define the family as a type, i.e., an institutionalized family system such as the American working middle class, male dominated, nuclear family; or the lower class, female centered family.

The philosophical ramifications of this deduction framework as applied to either of these polar-opposite family systems are not hard to imagine: 1. Family systems, including the non-working poor, are rational systems with specific goals. 2. One of the most basic, single purpose goals is to produce income. Whether this income is wild animal and plant nutrients; domestic foods; trade items; cash earned, inherited or doled; or some combination makes little difference. 3. In the pursuit of family production goals, this economic unit may be said to have a "core technology" including a cognitive map of its effective environment, a technical language, decision rules, production requirements, role differentiation, and a rationale. 4. Given the overwhelming importance of production goals and core technologies (adaptive strategies), family systems, as industrial corporations, tend to oppose changes that will disrupt the internal logic (principles of effective behavior) of the core technology. Whether this technology has been organized around a professional career, public assistance, or famine relief makes little difference.(13) 5. The deterioration of the family as an economic (adaptive) unit can be said to occur when the rational pursuit of *organizational* goals deteriorates to the rational pursuit of *individual* goals. Then the basic economic unit must be characterized some other way—for instance social networks. 6. Finally, contrasting characteristics of the lower and the working middle class are behavioral indicators of contrasts in their *effective economic environments*. Thus high achievement motivation, future time perspective, and internal locus of control—classic middle class correlates of success—are maladaptive in the economic environment operating with respect to the lower class. Moreover, public policies and programs aimed primarily at transforming their "undesirable" traits in the absence of goals to restructure their effective environments are disbenefiting the poor at the public's expense.(14)

SOCIAL MARGINALIZATION: AN INTERACTIVE PROCESS

Social marginalization is the interactive process of unequal environmental differentiation and adaptation which tends irreversibly to segregate people into economically peripheral, dependent, and counter productive modes of activity.

The dynamics of social marginalization originate in the labor market which is the core system for the distribution of economic and social benefits. Socialists like to contrast what they consider to be the central tenet of their system of distribution—"to each according to his need"—with what they say is the central tenet of the capitalist system—namely, "to each according to his ability." Actually, the American system adheres more to the practice of "to each according to his credentials and those without credentials will be administered benefits according to their worthiness." For the persistent moral and political dilemma of the capitalist model is the issue of the disposal of the labor surplus upon which its system of human resource utilization is predicated.

Independent of public regulatory bureaucracy, the labor market could not ration occupational statuses. It is merely one component, albeit a central one, in an enormous human resource processing system. The system can be depicted as a triad with the labor market occupying a fourth coordinant in the center of the triad (see Figure 1). At the top, public education starts the general flow of human capital with a set of administered credentials, predetermined and rationed according to the cultural preferences of the tax paying participants (principal investors) in the labor market system of benefit-rationing. Those not assimilated in the mainstream labor market are not summarily rejected outright, but, rather undergo a *process* of rejection involving underemployment, unemployment, exploitation (where returns to the employee are not sufficient to provide human investment capital), etc. The labor market overflow divides into two streams—one, the worthy poor flowing through the public welfare system; the other, the unworthy poor, who because of the risks associated with illegal economics, tend to flow through the criminal justice system. The components in the system are also connected by a network of information feedback.

Thus, marginalization is the consequence of: 1. The *cooperation* of a dual system of processing human resources in the system at large and most especially in the public school system. The educational establishment, although glibly

Figure 1

Human Resource Processing System

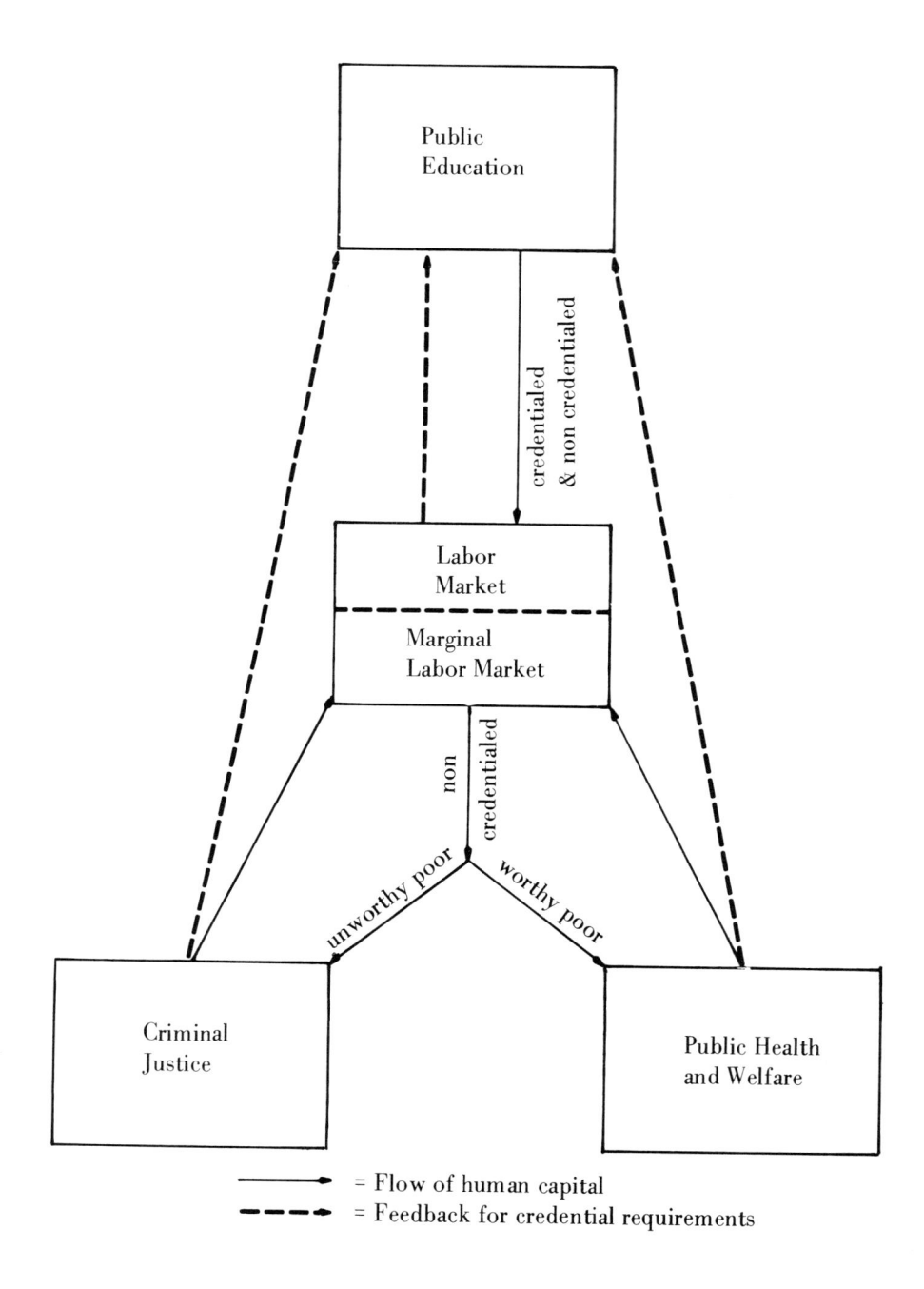

```
                    ┌──────────────┐
                    │   Public     │
                    │  Education   │
                    └──────────────┘
                           │
                    credentialed
                    & non credentialed
                           ↓
                    ┌──────────────┐
                    │    Labor     │
                    │   Market     │
                    ├──────────────┤
                    │  Marginal    │
                    │ Labor Market │
                    └──────────────┘
                     non  │  credentialed
                  unworthy poor  worthy poor
         ┌──────────────┐        ┌──────────────┐
         │   Criminal   │        │ Public Health│
         │   Justice    │        │  and Welfare │
         └──────────────┘        └──────────────┘
```

──────▶ = Flow of human capital
- - - -▶ = Feedback for credential requirements

promoting legally sanctioned mass education and the slogan that education is the solution to every human resource problem, was sociologically naive to the profound dilemma mass education would bring. Either public education would have to change other human resource institutions thereby fundamentally changing the entire human resource system, or it would have to develop subtle, infrastructural changes to maintain its integrity with the system at large. The result was the evolution of a cost/effective environmental system that can best be dichotomized as investment/caretaker with enormous incentive differentials. 2. Inter-institutional *cooperation* within the human resource system, i.e., each institution or component tends to distribute discriminatory information generally, thus the information stored with respect to the individual tends not only to be accumulative but consistent, thereby structuring environmental variability with respect to *classes* of individuals or human resources, system-wide. 3. And finally the *cooperation* of adaptive behavior of the social units with the structured environment they are experiencing—i.e., responding rationally to a net disincentive to develop competencies (invest) in the mainstream economic environment and a net incentive to develop competencies in exploiting marginal economic investments.

Thus the overall effect of time on under-investment in human resources in terms of both the individual and the group is irreversibility—namely, the longer under-investment continues, the more likely it is to continue.

What I have attempted to develop to this point are the essential components and conceptual boundaries of a dynamic system of interaction consisting of an effective economic environment on the one hand and adaptive social units as economic institutions on the other. However, we need time/space specifics beyond those implicit in the natural end-state phenomenon compelling public concern and scientific attention—namely, the urban ghetto. If the ghetto is the end product of a process, then it must have beginning and intermediate states with time/space specificity.

In social evolution generally, prior time implies prior space. In dealing with the evolution of human resource systems, especially in urban staging areas and ghettos, prior time and space implies a rural episode. Rural human resource systems vary with the core economy and the regional industrial system. In agricultural industries developed in the South, Southwest, and plantation Pacific—such as cotton, sugar cane, fruit crops, and vegetables, based upon labor-intensive technologies—labor markets are predicated upon the prevalence and maintenance of human capital under-investment—i.e., political and economic subordination. Rural institutions in this regard are well-known—i.e., race and class discrimination as a basis for differential administration of benefits in all key components of the human resource system including labor market, public education, criminal justice, and welfare. Although I have indicated that a movement to rural space is a movement back in time, I do not mean to infer that these institutions are extinct—quite the opposite.(15)

In the case of other rural industries like mining, wood products, cattle, and capital intensive agriculture, human resource systems tend to dictate human capital under-investment which is offset by growth cycles, but trapped into economic *cul-de-sacs* by natural resource decline or technological displacement.

The essential elements of a comprehensive rural/urban human resource system would include demographic linkages as a consequence of inequalities among all components of the two sectors. In cities the marginal and mainstream labor markets and public schools are more highly developed, in addition to which there are generally substantially greater benefits to urban public health and welfare. The net human resource flow then would be the youthful highly credentialed and youthful non-credentialed but non-urbanized—i.e., socialized in marginal labor markets and criminal justice and public welfare institutions.(16)

The rural community, in my view, offers a prime opportunity to develop a desperately needed comprehensive understanding of the human resource systems in urban industrial society, particularly with reference to social marginalization. In fact, putting it in the negative, without theoretical models and systematic research encompassing the human resource processes occurring in the rural economic setting, urban models are doomed to failure.

For this reason and consistent with its rural development mission, the Western Rural Development Center is conducting cross cultural (including non-white subordinant minorities), cross industrial investigations into rural human resource systems in five widely separated declining communities.(17)

Initially we hypothesized the declining community, in terms of family units and social networks, as being in the midst of becoming marginalized. But *recently* unemployed human resources were not easy to find and behavioral indicators were not all that apparent among the relatively small percentage we could locate. As a result of this initial inductive experience and some theoretical refinements, we are beginning to conceptualize the rural and urban episodes in the marginalization process as two distinct but related *phases.* They are related demographically—i.e., in the flow of human resources; but they are distinctive in terms of the institutional mechanisms by means of which human capital under-investment is maintained.

In order to develop comprehensive rural/urban models, we must have a theory that addresses two questions: 1. What precise roles do rural human resource systems play in urban human capital deterioration? and 2. Why adaptive social units (families and social networks) in declining rural communities do not tend to develop full blown adaptive cultures of marginality? To do this, the emphasis must be on systems and institutions that invest in, process, utilize, and disutilize human resources as opposed to emphasis on the people adapting to these systems. And it must provide theoretically for these institutions, including adaptive systems to function in significantly different ways in different phases of the process.

As a tentative effort, I will outline the beginnings of a processual model hypothesizing four phases in the marginalization of human resources.

SOCIAL DEVELOPMENT PHASE

Since communities in the *complete* economic and social sense generally arise as a result of an economy, it is hypothesized that every rural community had at least one such phase—i.e., when its core economy was developing, employment opportunity generally expanding and basic social systems were consonant with the core economy, and the core economy was adequate to sustain the social system—i.e., the economy was able to absorb the normal human capital flow generated within the community. If it were useful, one could hypothesize yet a prior phase to this called a "boom phase" characterized by extremely high returns to all forms of investment.

THE ATTENUATION PHASE

The dying community can be described as a prolonged state of imbalance between locally generated human capital and the local labor market. The primary effects of local economic depression are well known. However, it is the subtle, long term effects of this which should interest social scientists. For as long as the community produces children, its local human resource institutions must continue to operate. And the effects are not only felt locally but wherever its human capital flows.

Some of the universal behavioral tendencies of declining communities which lead to some interesting questions are: fear of outsiders; factionalism; narrowing of range or options of social expression to extremes—i.e., the church/bar syndrome; developing subtly discriminatory "credentials" for local labor markets; intensive cultural identification with occupational roles; and higher than average incidence of homicide, suicide, depression, and alcoholism.

These and other criterial characteristics of a declining community are less interesting than the interactive adjustment processes occurring within the human resource system and the long run effects this is having on the social and economic competencies of its people. The operating components in this situation are: Migration, education, labor, and family and social networks.

A declining community may or may not be suffering a net population loss, but there are important age, sex, and cultural differentials operating with respect to demographic movement. The outflow tends to be age, sex, and class specific which leaves a higher proportion of females, old and young, and local industry-oriented or traditional skills.(18)

This tendency in the migration system must be offset by the core industry if it is to continue to operate with the surplus labor pool upon which its technology and labor market are predicated. On the other hand, outmigration of the young is of central concern to other local institutions where social well being and self-interest are also threatened—namely, the family and school. The result is an enormous dilemma for public administration to which there will tend to be an ultra-conservative solution. Educating youth for urban labor markets is equivalent to investing in someone else's community whereas "educating" youth for local labor markets is an investment in the local community.

> Depending on the time span of development, local institutions will become re-oriented to provide a continuing stream of labor into the extractive industry. This influence pervades both formal (e.g., local government, especially schools) and informal (e.g., intergenerational work patterns) institutions, and is thus highly resistant to change. Moreover, the surplus of labor can be alleviated only slightly by the less dominant sectors of these specialized economies.(19)

Whether local human resources are destined for migration or local utilization, this kind of educational policy is tantamount to creating a *general* set of technical skills juxtaposed with a highly specialized, culture-specific set of social skills designed to protect access to declining local labor markets. But, regardless of the culture-specificity operating with respect to job rationing in the local labor markets or credential administration in the local schools, the net effect on human capital is compound. Human capital investment declines and institutional capacities to create human capital decline thus insuring a continuing transfer of human capital deficit to other human resource systems in the national economy. Thus the concept of *setup* is applied to the declining or economically attenuating rural community.

In other respects as well, the concept of *setup* is appropriate because the consequences of human capital under-investment do not tend to be felt in the local setting, perhaps primarily because the functional role the family is assigned with respect to other human resource institutions is not diminutive. It tends to have institutional integrity, its assets are capable relative to local demands, and its wisdom is generally valid. Putting it differently, the rural economic/social setting does not constitute the effective environment for social marginalization; but nevertheless, it plays a decisive role.

THE PRECIPITATION PHASE

The staging area concept applies to a dramatic change in the economic environment of the declining community which invalidates the economic adjustment patterns of the adaptive social units. However, it means more than simply a change in the environment, it means a particular sequence of economic environments.

A declining community in terms of human resources is under-investing in human capital at a gradually increasing rate, which theoretically makes them susceptible to social marginalization under certain conditions. Thus a rural family socialized in a declining community and still living and underemployed in that community will experience a precipitation phase by urban flooding of the local setting as well as by immigrating to an urban area. In either case, staging is a forced decision situation—one where self-conscious change is unavoidable because it tends to precipitate the obsolescence implicit in the old human resource system. It brings people into an abrupt encounter with extensive social environmental variation.

Decline of the importance of the extended family, perhaps even the nuclear family, is inevitable not because urbanization is synonymous with social decay, but because of the greater emphasis on public institutions for the processing and regulation of human resources in urban industrial systems. Thus new institutions develop which compete with the old in human capital investment functions. Moreover, in terms of individual strategies, other adaptive units—social networks, etc.—may be more useful in this quest for economic survival.

Economic staging environments also offer a variety of niches for exploitation. This milieu provides other markets and incentives to develop new skills as well as other human resource investment institutions. However, before one gets carried away with economic growth cliches, it is well to be reminded that in the modern, regulated environments and controlled labor markets of the 20th century as opposed to the relatively unstructured social environments of the 19th century frontier, non-credentialed human resources are likely to find their greatest assets to be in marginal markets.

THE CLOSED SUBSYSTEM PHASE

The ghetto is to be regarded as a subsystem because it is a natural sector of human resources systems in Western industrial societies. It is closed because it is an environmental system with behavioral specificity antithetical to the main system. That is to say the competencies necessary to behave appropriately in its institutions, such as welfare and criminal justice, compel inappropriate behavior in the labor market. Moreover, the human capital accounting process records these experiences as negative credentials, thereby reinforcing the environmental specificity of the subsystem.

Economically, the ghetto constitutes a subeconomy in that it has specialized markets for goods and services, the exploitation of which again requires specialized competencies that must be learned in specialized human capital institutions—e.g., gangs and other social networks. The problem is that these institutions also socialize their members within an affective and ethical system antithetical in many respects to mainstream society thereby increasing conflict and frustration when the individual attempts to translate.

Hence, three, as opposed to two, economic sectors having human capital implications are revealed in this time/space model: rural, urban, and ghetto.

CONCLUSION

In summary, inequality within the framework of marginalization theory is seen as the consequence of artificial habitats. These artificial habitats are created by the cooperation of institutional mechanisms designed to solve the dilemma of labor surplus while protecting class privileges. To achieve this solution, there has been and continues to be the systematic cooperation of private economic institutions and public human resource institutions; hence, the concepts of effective economic environments and adaptive behavior of individuals and families as rational strategies for the exploitation of these environments.

In order to understand the role of public human resource institutions in the maintenance of unequal habitats relative to human capital formation, they must be seen as single institutions with dual functions creating dual environments. In this model, public education is but one component of several major components using short run decision rules to invest in the human resources showing the highest returns and to regulate the remainder.

In order to complete the model of the marginalization process time/space relationships must be comprehended. A partial theory of how this functions is represented in four phases having time/space implications with the rural community human resource system revolving around a core industry. The decline of the core industry sets the stage for human capital obsolescence and subsequent specialization in marginal subeconomies in urban settings.

Finally, the concept of inequality must be applied to social habitats and not to the rational (adaptive) system of behavior specific to those habitats. To compare behavioral characteristics cross-habitationally irrespective of their environmental appropriateness is grossly unscientific. Most concepts of intelligence, including IQ, presume a universally uniform habitat. Either habitat specificity must be explicitly recognized in these concepts, or new concepts of competence should be developed based upon environmental, particularly institutional, relativity. Perhaps when this is done, it will be possible for social policy to imagine more socially valid definitions of the problem of human disability to perform in the American mainstream.

NOTES

(1) Roy E. Rickson, "Industry, Science and Pollution: Some Problems that Industrial Societies have in Developing a Quality Environment." Dept. of Sociology, University of Minnesota, St. Paul, mimeograph, 1974.

(2) Charles C. Gillispie, "Mertonian Theses," Book Review of *The Sociology of Science* Theoretical and Empirical Investigations. Robert K. Merton, *Science*, Vol. 184, 1974, p. 656.

(3) For comprehensive critiques see Charles A. Valentine, *Culture and Poverty*. University of Chicago Press, 1968; and Eleanor Burke Leacock, Editor, *The Culture of Poverty: A Critique*. Simon and Schuster, New York, 1971.

(4) For example see Daniel Patrick Moynihan, "The Negro Family: The Case for National Action," *The Moynihan Report and the Politics of Controversy*. M.I.T. Press, Rainwater and Yancey, editors, 1965.

(5) Gutorm Gjessing, "The Social Responsibility of the Social Scientist." *Current Anthropology*. Vol. 9, No. 5, December, 1968, pp. 397-402; and "Commentary" and "Book Reviews" sections, *Human Organization*. Vol. 31, No. 1, Spring, 1972, pp. 95-110.

(6) Milton M. Gordon, *Assimilation in American Life: The Role of Race, Religion, and National Origins*. Oxford University Press, New York, 1964, pp. 68-70.

(7) Nathan Glazer and D.P. Moynihan, *Beyond the Melting Pot: The Negroes, Puerto Ricans, Jews, Italians, and Irish of New York City.* M.I.T. Press and Harvard University Press, Cambridge, 1963.

(8) "The culture of marginality" is more precise than "the culture of poverty" since sheer material (including dietary) deprivation is less at issue than the deprivation of economic and social status, which may include income usually more than adequate for physical survival. For a case study of institutionalized starvation that could be termed "the culture of poverty" see Colin Turnbull, *The Mountain People.* Simon and Schuster Inc., New York, 1972.

(9) Octavia I. Romano-V, "Institutions in Modern Society: Caretakers and Subjects," *Science*, Vol. 183, New York, February 22, 1974, pp. 722-725.

(10) Harland Padfield and Roy Williams, *Stay Where You Were: A Study of Unemployables in Industry.* J.B. Lippincott Co., Philadelphia, 1973.

(11) Economic studies of occupational performance of the lower class in terms of incentives/disincentives, although relatively infrequent, tend generally to support the rational theory of motivation. For examples see Paul Gayer and Robert S. Goldfarb, "Job Search, the Duration of Unemployment, and the Phillips Curve: Comment and Reply," *The American Economic Review.* Vol. LXII, No. 4, September, 1972; Bennett Harison, "Education and Underemployment in the Urban Ghetto." *The American Economic Review.* Vol. LXII, No. 5, December, 1972; Leonard Weiss and Jeffry G. Williamson, "Black Education, Earnings, and Inter-regional Migration: Some New Evidence," *The American Economic Review.* Vol. LXII, No. 3, June, 1972; and Burt Shlensky, "Evaluation of Training Programs for the Disadvantaged Through a Psychological Cost-Benefit Model," Hadley C. Ford & Assoc., 6 East 43rd Street, New York, Mimeograph, 1974.

(12) Rickson, *op. cit.*

(13) For an interesting instance of the social and economic ramifications of adaptation to famine relief see Turnbull, *op. cit.*, pp. 280-282.

(14) For empirical evidence in support of this statement see Theodore D. Graves, "Urban Indian Personality and the 'Culture of Poverty,'" *American Ethnologist.* Vol. 1, No. 1, February, 1974, pp. 65-86; and Padfield and Williams, *op. cit.*, pp. 205-253.

(15) Harland Padfield and William E. Martin, *Farmers, Workers and Machines: Technological and Social Change in Farm Industries of Arizona.* The University of Arizona Press, Tucson, 1965; and Harland I. Padfield, "Agrarian Capitalists and Urban Proletariat—The Policy of Alienation in American Agriculture," *Food, Fiber and the Arid Lands.* The University of Arizona Press, Tucson, 1971, pp. 39-46.

(16) Weiss and Williamson, *op. cit.*, p. 379.

(17) The locations including names of universities and principal investigators are as follows: Eastern Washington, small businessmen in rural communities, Dr. Paul Barkley, Department of Agricultural Economics, WSU; Oregon, study of wood products industries workers facing layoff, Dr. Joe Stevens, Department of Agricultural Economics, OSU; California, small farm operators in Colusa County, Dr. Jerry Moles, Department of Anthropology, UC/Davis; Hawaii, sugar and pineapple workers facing layoff due to phasing out of these industries on the island of Molokai, Dr. Robert Anderson, Department of Agricultural Economics, UH; Arizona, workers facing layoff due to phasing out of copper mining

operations, Dr. William Martin, Department of Agricultural Economics, UA. For more information on this project please write to the author at the address indicated.

(18) Lloyd D. Bender, Bernal L. Green, and Rex Campbell, "Rural Poverty Ghettoization," Dept. of Agricultural Economics & Economics, Montana State University, Bozeman, mimeograph, 1971.

(19) Joe B. Stevens, "On the Process and Consequences of Job Rationing in Declining Extractive Industries." Abstract, Dept. of Agricultural Economics, Oregon State University, Corvallis, mimeograph, 1974.

FALSE CORRELATIONS= REAL DEATHS: THE GREAT PELLAGRA COVER-UP, 1914-1933

Allan Chase

At the turn of the twentieth century, the white, Anglo-Saxon, Protestant (WASP) population of the Southern United States was categorized by the American eugenicists and other scientific racists as a sub-race of genetic paupers.

It is a cardinal tenet of the eugenics dogma "that there is in existence a definite race of chronic pauper stocks."(1) In the Northern states, the poor white members of this "definite race of chronic pauper stocks" were often labeled (or libeled) as Jukes and Kallikak types. In the South, the native white poor were solemnly described to the U.S. Supreme Court by Harry Hamilton Laughlin, Supervisor of the Eugenics Record Office and co-editor of its *Eugenical News*, as being, by biological inheritance, "The shiftless, ignorant, worthless class of anti-social whites of the South."(2)

The then powerful American eugenics movement had, by 1914, succeeded in making these subjective and racist judgments appear to be tested scientific facts to the college-educated lawyers and legislators and presidents who wrote, enacted, and administered all public policies.

Thus, long before the era of I.Q. testing, pseudogenetics and pseudobiology were already being used with devastating effect to prove that—by genetic endowment—the WASP poor of America in general, and of the South in particular, were doomed by their "inferior blood" to be the behavioral, intellectual, and economic inferiors of the non-poor WASPS who came of "superior breeding stock."

The eugenical myth of the poor whites of the South being hereditary white trash was severely shaken by the work of two United States Public Health

Service scientific workers between 1900 and 1914. They were the parasitologist Charles Wardell Stiles, son and grandson of Methodist ministers in Connecticut—and the epidemiologist, Joseph Goldberger, brought from his native Hungarian village to the lower East Side of New York at the age of six by his poor Orthodox Jewish immigrant parents.

It was Stiles who not only proved that hookworm infection existed in the United States, but also that it was endemic among the poor people of the South. As early as 1903, *The Popular Science Monthly* reported Stiles as noting that:

> One of the most important symptoms of 'hookworm' disease is an extreme lassitude, both mental and physical; this condition is due to the emaciation and to the watery character of the blood, which does not properly nourish either the brain or the muscles. Now, curiously enough, it is especially in the sandy areas of the South that the poorer whites, known as 'poor white trash' are found.... He (Stiles) states that if we were to place the strongest class of men and women in the country in the conditions of infection under which the poorer whites are living, they would within a generation or two deteriorate to the same poverty of mind, body and worldly goods which is proverbial for the 'poor white trash.'(3)

One of the most important of Stiles' findings during his field investigations in the South was that

> hookworm disease is especially prevalent among children, and that it not only interferes with their school attendance, but that children who are afflicted with the malady and who have gone from the sandy districts to a city have the reputation among their teachers of being more or less backward and even stupid in their studies.(4)

As early as 1902, Stiles was telling all who would listen that hookworm disease could easily be prevented. All that had to be done, Stiles wrote, was to build privies or toilets for all families in those areas of moist, sandy soil where poor people traditionally walked in their bare feet.

There are, of course, two other ways to protect people from being infected by the hookworm larvae present in the feces of hookworm victims—or, if they become infested with the parasites, to keep the wormy nuisances from becoming pathogens. These prophylactic modalities are: 1. to make sure that all people who walk in hookworm-infested human excrement wear strong, larvae-proof shoes; and 2. to make certain that all poor people harboring hungry, blood-sucking hookworms in their guts get enough protein- and iron-rich foods to meet both their own biological needs and, at the same time, also satisfy the nutritional requirements of the parasitic hookworms that live on the blood of their human hosts.

Hookworm, in short, is an eminently preventable disease of poverty.

In December, 1902, Irving C. Norwood, a reporter for the New York *Sun*, attended a lecture by Stiles and reported that the "Germ of Laziness" had at last been discovered. This catchy concept not only stirred the imagination of the public. It also helped, in time, to convince John D. Rockefeller, Sr., that the moment had come to spend upward of one million dollars on an "aggressive campaign" against this environmental cause of the chronic "laziness" and low economic productivity of poor Southern people.

Between 1909 and the end of 1914, the Rockefeller commissions spent $797,888.36, and had cooperated with scores of Southern state and county health departments in the examination of nearly 1.3 million people, and in the treatment of 700,000 victims of hookworm disease. Some 39% of all Southern school children, most of them white, were found to have hookworm disease, "with a variation from one county to another of from 2.5 to 95%."(5)

Many of the 700,000 Southern victims of hookworm infection—and of the body-weakening and mind-crippling nutritional anemia that hookworm disease causes—were also, it soon developed, sufferers from an equally ubiquitous chronic degenerative disease. This second disease was called pellagra.

The word pellagra was from the Italian words for "angry skin." The disease caused widespread body rashes; followed by lassitude, loss of appetite, weakness, and irritability. Pellagrins who survived these early symptoms progressed to inflammation of the mucous membranes and swollen tongues and such pains as to make even a sip of water too agonizing to bear. Finally, the minds of pellagrins progressed from disorientation to delirium to hallucination. Pellagra, in short, drives its victims mad.

Like hookworm infection, the incidence of pellagra in the United States was not recognized, clinically, until after the turn of the century—when it turned out that upward of a third of the patients in our county madhouses proved to be pellagrins. By 1908, a National Association for the Study of Pellagra had been organized. In 1909, a National Conference on Pellagra was held in Columbia, South Carolina at the State Hospital for the Insane.

Pellagra not only led to insanity. Endemic pellagra, along with hookworm, roundworm, and other energy-diminishing disorders, acted to discourage Northern manufacturers from taking advantage of low wage scales and other regional inducements to build their textile mills and other factories in the South. Pellagra also threatened our export markets for corn, because of the then widely-held belief that the pellagra endemic in the South was caused by "spoiled" corn.

A national Pellagra Commission, sponsored by the philanthropists Thompson and McFadden, was organized in 1913 by the New York Post-Graduate Medical School to discover the cause and cure for the disease. The senior investigators of the Pellagra Commission—J.F. Siler, an Army doctor, and P.E. Garrison, a Navy doctor—were first rate workers. Garrison and Siler shared the hypothesis that pellagra was an infectious disease, and they pointed their field studies toward the isolation of the hypothetical pellagra pathogen. It was a perfectly legitimate and logical hypothesis. It merely happened to be the wrong hypothesis for this disease.

In 1914, Joseph Goldberger, who had already won recognition as the leading field investigator of infectious diseases in the United States Public Health Service, was second in command of Stiles' laboratory in Washington, D.C. When the government assigned him to the task of organizing and heading a task force to work on pellagra, Goldberger first tried and failed to induce pellagra in monkeys by innoculating them with biological materials from pellagrins.

On his initial field trip to the South, Goldberger saw enough to suggest to him that pellagra was not an infectious disease but a disorder of poverty. His team of forty-four pellagra investigators fortunately included a very talented economist, Edgar Sydenstricker. Goldberger put him to work at "tracking down economic data on labor, family budgets, family dietaries, food prices."(6)

As we know, Goldberger and his Public Health Service field investigators proved that, beyond all doubt, pellagra was an ordinary deficiency disease caused by the lack of foods containing what Goldberger designated, originally, as the Pellagra Preventing (PP) factor. PP was later identified as niacin, one of the B-vitamins, and for a short time was also called Vitamin G, in honor of Goldberger. The meats, poultry, dairy products, fruits and vegetables containing niacin (or its metabolic precursor, tryptophan) were far more costly than the corn that was the staple of the diet of the Southern poor.

As Sydenstricker reported, retail food prices had climbed 60% since 1900, but wages had not risen by more than 25% during the same period, "while in many industries and instances there has been an increase of less than 5% since 1907 and 1908 in the South."(7) Pellagra-preventing foodstuffs were simply too expensive for the families stricken with one or more cases of chronic pellagra.

The publication, in various Public Health Service reports and clinical journals, of Goldberger's discoveries on how to cure, prevent, and even cause pellagra created a scientific sensation.(8) Garrison and Siler, pausing only to congratulate Goldberger for solving the riddle of pellagra, promptly resigned from the Pellagra Commission and went to other assignments. Goldberger was invited to deliver Harvard University's prestigious Cutter Lecture in Biology for 1915. Congratulations poured in on Goldberger, and the United States Public Health Service, from medical leaders around the world. Joseph Goldberger had not only conquered pellagra; with his pellagra research, he had also founded the new medical specialty of clinical nutrition.(9)

In the best of all possible Panglossian worlds, this triumph should have meant that the state and federal governments would then join forces to wipe out pellagra and hookworm in the South as thoroughly as the construction of clean water systems had wiped out cholera in England in the decades after Dr. John Snow, at the Broad Street pump, showed dirty drinking water to be associated with London's cholera outbreaks. America, however, was a very different place at a different time.

In 1914, the year Goldberger showed pellagra to be a non-hereditary disease of malnutrition, there were 847 reported deaths from pellagra in the United States. Between 1914 and 1928, the year of Goldberger's untimely death, pellagra mortalities had soared to a total of 6823—or eight times the number of reported pellagra deaths during the year in which, fourteen years earlier, Goldberger had published the first of his famous reports on pellagra. By 1929, government records showed some 200,000 cases of pellagra reported in the nation, including some 6623 pellagra deaths.

These pellagra mortality and morbidity figures represented only a fraction of the actual totals. By definition, a clinical diagnosis calls for the presence of both a patient *and a physician;* pellagra, as indeed all other poverty diseases, is always a grossly under-reported disease.

This triumph of entropy over epidemiology did not occur in a social or an historical vacuum. It happened because, like Stiles' prior work on hookworm, Goldberger's work on pellagra threatened to destroy the American eugenics movement, the central dogma of which had all mental diseases and behavioral disorders classified as being at least 80 to 100% hereditary.

There was nothing the eugenics leaders could do about trying to refute Stiles and his discoveries about the psychopathologies of hookworm disease and its side-effects. To attack the "Germ of Laziness" answer to the question of what made the South's 100% Nordic "poor white trash" lazy, shiftless, stupid, and, above all other human sins, *poor,* was now, in effect, to attack the Rockefeller Foundation and the Rockefellers themselves. To attack the Rockefellers was a folly that not even the most hard-core eugenicists—tigers though they might be against the poor whites, against the Jews, and against the non-Nordic immigrants—were about to undertake.

Goldberger's findings that pellagra, and the mental disease it caused, were neither of them genetic were targets more in keeping with the courage of the leaders of the eugenics movement. Because of the etiological links between madness and pellagra, Charles Benedict Davenport, the Director of the Eugenics Record Office, had as early as 1913 become involved with the Pellagra Commission. When Goldberger made the work of the Commission superfluous in 1914, causing Garrison and Siler to leave it, Davenport took control of its work and the preparation of its final report, "Pellagra III: Third Report of the Robert M. Thompson Pellagra Commission of the New York Post-Graduate Medical School and Hospital."

Two years *after* Goldberger's first report on the real causes of pellagra, Davenport, who was a zoologist with no clinical training or experience, and Elizabeth Muncey, a physician on the payroll of Davenport's institute, published two articles in the July 1916 issue of *The Archives of Internal Medicine.*(10)

During the same month, both articles—"The Hereditary Factor in Pellagra," by C.B. Davenport, Ph.D., and "A Study of the Heredity of Pellagra in Spartanburg County, South Carolina," by Elizabeth B. Muncey, M.D.—were reprinted as Bulletin No. 16 of the Eugenics Record Office. When the final report of the Pellagra Commission was published in 1917, the same pair of articles on pellagra as an hereditary disease of people of inferior breeding stock materialized for a third time as the closing two chapters of this long and very authoritative report.

Davenport's thesis was very simple. Pellagra, he wrote, was "the reaction of the individual to the poisons elaborated in the body, probably by a parasitic organism. His thesis accords with the conclusion of Siler, Garrison (sic), and Macneal that pellagra is in all probability a specific infectious disease communicable from person to person."

By March 16, 1916, when Davenport submitted this paper to the *Archives of Internal Medicine*, he *knew* that Garrison and Siler had long since revoked their former infection hypothesis (it was never a "conclusion") in the face of Goldberger's elegant biological proofs that it was incorrect. Davenport also knew very well that Garrison and Siler both agreed, with Goldberger and his Public Health Service colleagues, that pellagra was a simple and predictable sequela of hunger—a classic poverty disease.

Nevertheless, with a Galtonian disdain for the mere facts of human biology, Davenport continued to label pellagra as an infectious disease. More than that, he also insisted that

> in the pellagra reaction (to the mysterious 'pellagra germ') there is a hereditary factor. . . . if there is one thing of which experience perfectly assures us it is that different individuals react dissimilarly to the same stimulus. Such dissimilarity of reaction is conditioned both by fundamental dissimilarity in the constitution of the organism and by dissimilarity in antecedent experiences of the organism; but the latter, in turn, is conditioned in part by the former; so that the fundamental dissimilarity of the constitution of the organism must be held to be the principal cause of the diversity which persons show in their reaction to the same disease-inciting factors.
>
> This constitution of the organism is a racial, that is, hereditary factor. And if it appears that certain races or blood lines react in pellagra families in a specific and differential fashion, that will go far to prove the presence of a hereditary factor in pellagra. . . . colored persons, who differ from most white people in having more or less black blood, are less subject on the whole to the disease than white persons.

Davenport's conclusions were in the true eugenic traditions:

> Pellagra is not an inheritable disease in the sense that brown eye color is inheritable. The course of the disease does depend, however, on certain constitutional, inheritable traits of the affected individual. . . . Pellagra is probably communicable, but how the communicated 'germ of the disease' shall progress in the body depends, in part, upon constitutional factors. . . . When both parents are susceptible (sic) to the disease, at least 40 percent, probably not far from 50%, of their children are susceptible; an enormous rate of incidence in a disease that affects less than 1 percent of the population on the average. . . .

In the final report of the Pellagra Commission in 1917, Goldberger's work was, except for one footnote on pages 226-227, completely ignored. This footnote dismissed, as a failure, Goldberger's famous experiment that disproved the transmissability of pellagra. The footnote stated:

> The relative insusceptibility of pellagra of young adult men is generally recognized. If one should desire, therefore, to obtain positive results in experimental inoculations of man, it would obviously be wise to select the experimental subjects from some other group of the population, children from 2 to 10 years of age and childbearing women from 20 to 45 being most valuable for this purpose. The recently published negative experiment carried out by the United States Public Health Service (Goldberger, Joseph: The Transmissibility of Pellagra, Pub. Health Rep., 1916, 31, 3159) would appear, therefore, to have little or no bearing on the question of the transmissibility of pellagra, although it may be of some value in confirming the previously well known fact that healthy men in the more active period of life are quite refractory to the disease.

Goldberger was, according to Davenport, ignorant of the non-fact that "the relative insusceptibility to pellagra of young adult men is generally recognized."(11)

Articles that Garrison and Siler had previously published in various issues of the *Archives of Internal Medicine,* and which both authors agreed had been contradicted by Goldberger's findings, were now republished as originally written in the 1917 final report of the Pellagra Commission. These articles by Garrison and Siler constituted the majority of the chapters in the final, 1917 report of the Pellagra Commission. The only concession to reality and truth made by Davenport and his co-editors was a footnote on the bottom of each of the chapters signed by Garrison and Siler. This bit of fine print informed anyone who could read type that small, that:

> the final copy of the paper itself has been written since Dr. Garrison and Dr. Siler were recalled to active service in the Medical Corps, U.S. Navy, and the Medical Corps, U.S. Army, respectively. They are, therefore, not responsible for the observations of the last two years, for the compilation of the data, or for the deductions drawn from them.

Typical of what Goldberger denounced, in exasperation, as the "false correlations" between the sewage of poor neighborhoods and/or the genealogy of pellagrins with the real causes of pellagra was Table 5 in Chapter VII of the final report of the Pellagra Commission. It proved only that poor people are as prone to have the diseases of hunger as were their equally poor neighbors, relatives, and ancestors:

RELATIONSHIP OF PELLAGRINS IN FAMILIES WITH PELLAGRA IN THE THIRD GENERATION

1 Grandmother (1911-1913)		1 Granddaughter (1914)	Direct
2 Grandmothers (1893-1911) (1910-1913)		1 Grandson (1913)	Direct
1 Grandfather (1912)	Mother (1913-1914)	2 Granddaughters (1913-1914)	Direct
1 Grandfather (1909)	[Grandmother] (1911)	2 Granddaughters (1913)	Direct
1 Grandfather (1912-1913)	Father (1912-1913)	1 Grandson (1911-1912)	Direct
1 Grandfather (1901)	Mother (1907-1913)	1 Granddaughter (1911-1913)	Direct
1 Grandmother (1910-1912)	Mother (1909-1913)	2 Grandsons (1910-1913)	Direct
1 Grandfather (1908)	Mother (1904-1913)	2 Grandsons (1912)	Direct
1 Grandmother (1910-1912)	Mother (1909-1913)	2 Grandsons (1910-1912) (1913)	Direct
1 Grandmother (1900)	2 Mothers (1905) (1910-1911)	2 Grandsons (1911-1912)	Direct
1 Grandmother (1910-1914)	4 Daughters (1910) (1913) (1914)	2 Granddaughters (1914)	Direct
1 Grandfather (1913)	Son-In-Law (1912-1913)	1 Granddaughter (1914)	Direct and Indirect
1 Grandfather (1910-1912)	Daughter-In-Law (1912-1913)	2 Grandsons (1913) (1914)	Direct and Indirect
1 Grandfather (1900)	Daughter and Son-In-Law (1911-1913) (1912)	1 Grandson (1911)	Direct and Indirect

| 1 Grandmother (1910) | Son and Daughter-In-Law (1910-1912) ((1910) | 3 Grandchildren (1912) | Direct and Indirect |

There are also

| 1 Step-Grandfather (1911-1914) | 3 Stepchildren (1910) (1910) (1913) | 2 Step-Grandchildren (1910) (1913) | |

| 1 Step-Grandmother (1912) | 1 Stepdaughter (1910-1914) | 3 Step-Grandchildren (1912) | |

Dr. Munsey's chapter in "Pellagra III" was liberally seasoned with batches of the famous pedigree trees so dear to the hearts of eugenic scholarship, then and now. One particular pedigree chart traced the prevalence of pellagra and other poverty diseases such as "hookworm disease followed by nervous exhaustion and debility" in the family tree and personal history of "Pellagrin No. 25."(p. 429) This poor woman died in a madhouse after having had hookworm disease in 1908, and four attacks of pellagra-followed-by-madness, in 1914. The pedigree chart showed that this woman had three children, two of whom were pellagrins.

Whether Davenport really believed that he and Muncey had destroyed the scientific validity of Goldberger's findings is, today, of little moment. All that does matter, in history, is that in 1916 and 1917, Davenport and Muncey had told the American society what it wanted to hear about this endemic disease of poor people.

If pellagra was indeed as Davenport claimed, an infectious disease of genetically inferior white Anglo-Saxon Protestant breeding stock, then it could not possibly be either prevented or cured by such social actions as minimum wage laws that would enable the working poor of the South to earn enough money to provide their families with the meat, eggs, dairy products, and vegetables that Goldberger claimed would prevent and/or cure pellagra. Nor could pellagra be prevented, in the families of America's "definite race of chronic pauper stocks," by the societal provision of food supplements to families who were too poor to buy meat, fish, poultry and other sources of niacin.

It was cheaper, it seemed, to believe in the pseudogenetic myth of pellagra as an infectious disease of a sub-race of inferior hereditary stock. Such self-defeating socioeconomic illusions were, of course, among the primary reasons why the pseudogenetic, pseudoinfectious pellagra hypothesis—as detailed in the third and final report of the Pellagra Commission—was accepted as scientific truth by the lawyers and businessmen who then made and administered governmental priorities and policies concerning health, education, and human welfare.

The Davenport-fabricated pseudodescription of pellagra became so integral an aspect of American conventional wisdom that, as late as 1938, A.M. Stimson, the Medical Director of the U.S. Public Health Service, had to include a section on pellagra in his history of the service's bacteriological investigations. He wrote:

> It is known that there are some die-hards who still cling to the infectious theory of pellagra. Even if they grant some influence of diet, they propose an ultimate exciting cause among the micro-organisms. Some plausibility to these claims was offered in early days when individuals were found who, although affluent and able to command every table luxury, still had pellagra. Goldberger himself observed these cases until he found that, although the table might be laden with viands of all sorts, the patient, through some personal idiosyncrasy, selected the wrong ones for his personal consumption. It is up to these objectors to produce the causative organism, and also to explain why, in prosperous years, pellagra almost disappears, to return again with hard times and limited diets, or that more recent refinement of epidemiology which shows that even in hard times, when the price of cotton is so low as to make it not worth while to plant, the cow and the kitchen garden which takes its place entirely prevent pellagra.(12)

The state of knowledge of the American medical community in 1917 happened to be not the least of the other reasons why—in state, county, and city health departments and professional medical societies—Davenport's pseudoscientific pellagra hypothesis was able to prevail over Goldberger's testable scientific findings.

As of 1917, a *majority* of our American physicians were the graduates of inferior, commercial, scientifically, and clinically grossly inadequate medical schools. Most of these "medical schools," even those nominally attached to colleges and universities of high repute, were little more than diploma mills. Outside of Johns Hopkins, and one or two other schools that had joined the Hopkins in seeking to emulate the better and more scientifically-oriented European medical schools, the average American medical school had yet to catch up with nineteenth-century biomedical advances. As the President of the Carnegie Foundation for the Advancement of Teaching wrote in his foreword to the famous study of medical education in the United States and Canada made for the foundation by Abraham Flexner between 1908 and 1910:

> For the past 25 years there has been an enormous over-production of uneducated and ill trained medical practitioners . . . due in the main to the existence of a very large number of commercial schools.(13)

These "uneducated and ill-trained medical practitioners" were also, in 1917 and at least a dozen years that followed, the medical cohort from which the majority of the nation's municipal, county, state, and federal health officers were derived.

Thus, the lack of scientifically viable societal policies concerning the standards of medical education, and the clinical training of America's doctors, between 1880 and 1920, was to create a population of American physicians who were intellectually quite capable of accepting the impressive, 500-page report of the Robert M. Thompson Pellagra Commission of the New York Post-Graduate Medical School as the ultimate scientific conclusions on the nature, causes, and treatment of pellagra.

How costly this ignorance of enough biology to be able to distinguish between the scientific epidemiology of Goldberger and the pseudomicrobiology of Davenport was to this nation is seen in Table 1.

TABLE 1

DEATHS FROM PELLAGRA BY COLOR IN THE UNITED STATES

YEAR	TOTAL DEATHS	WHITE PEOPLE	ALL OTHER PEOPLE	HISTORICAL EVENTS AND SOCIAL CONDITIONS DURING SAME YEARS
1900	2			Most American doctors unaware of pellagra
1914	847			Goldberger discovers "that pellagra is a vitamin deficiency disease analogous to scurvy and beriberi."
1915	1058			Goldberger shows how to cause, cure, and prevent pellagra. Davenport publishes *The Hereditary Factor in Pellagra* in *The Archives of Internal Medicine*.
1916	1807			
1917	2843			*PELLAGRA III*, final report of Pellagra Commission, describes pellagra as an infectious disease affecting primarily white people genetically susceptible to its "germ."
1918	3126			
1919	2568			
1920	2122			
1921	2348			
1922	2514			
1923	2245	1143	1102	First year U.S. Public Health Service records pellagra deaths of non-whites. The black 10% of America's population prove to suffer 50% of its pellagra mortalities.

1924	2206	1086	1120	
1925	3049	1384	1665	
1926	3501	1724	1777	
1927	5091	2351	2740	
1928	6523	2689	3834	Death of Joseph Goldberger.
1929	6623	2781	3842	Stock market crash triggers decade of The Great Depression, impoverishing the nation.
1930	6106	2722	3384	
1934	3602	1914	1688	Start of Federal work and food relief in 1933 improves diets of poor families. Start of TVA provides many new jobs.
1935	3543	1963	1580	Cheap electric power from new TVA dams gives South new grain mills, new markets for local grain, new dairy and poultry industries, and lowers prices of niacin-bearing foods.
1936	3740	2129	1611	
1937	3258	1804	1454	
1938	3205	1707	1498	
1939	2419	1404	1015	Start of World War II in Europe creates new factories and jobs in the South.
1940	2123	1270	853	
1941	1836	1137	699	America enters World War II.
1968	15	12	3	During and after World War II, the diets of Southern poor are improved, as continuing industrialization of the South creates more jobs for white and non-white poor. Voting Rights Act of 1955 and other civil rights reforms opens doors to better jobs to Southern blacks.

(Mortality data from U.S.P.H.S. Center for Disease Control, Atlanta, Georgia)

In 1917, as today, the social and human costs of the general ignorance of science on the part of our educated classes were catastrophic.

The Great Pellagra Cover-Up of 1914-1933, which kept the medical benefits of Goldberger's work on pellagra from the entire nation for two decades, was the greatest triumph of scientific racism since 1826—when Thomas Malthus denounced as reprobates those doctors who devised specific "remedies for ravaging diseases; and those benevolent, but much mistaken men, who have thought they were doing a service to mankind by projecting schemes for the total extirpation of particular disorders."

As every doctor knows, had American society, acting on Goldberger's findings, totally extirpated pellagra by making certain that the wages of the poor allowed them to buy all of the foods needed to prevent pellagra, the same benevolent social action would have set off a series of clinical reactions that, at the same time and for the same investment, would have had equally salubrious affects on scores of other particular disorders of malnutrition. Such as, to name but five groups of the ravaging diseases and disorders of chronic malnutrition: 1. low birth weights; 2. inadequate growth levels during the "period of human development extending from the second trimester of gestation well into the second postnatal year, during which the brain appears to have a once-only opportunity to grow properly"—the period when the human brain achieves 75% of its ultimate mature weight; 3. the tragically high rates of often fatal infectious diseases, from measles and diphtheria to pneumonia and tuberculosis, that are always associated with chronic undernourishment the world over; 4. the growth-retarding effects of lifelong malnutrition on the pelvis of the human female, which makes cephalo-pelvic disproportion (CPD) a major cause of injuries during childbirth to the brains and bodies of the children of the world's poor; and 5. the staggeringly high incidences of infection-caused and often fatal dysenteries and diarrheas in the malnourished children of the poor.

It would not be overstating the seriousness of the clinical effects of the Great Pellagra Cover-Up launched by Davenport and the American eugenics movement he organized and led for over thirty years to say that, in terms of *preventable* morbidities and mortalities in only these five categories of common diseases and disorders known to be exacerbated by chronic malnutrition, they added up to millions of completely avoidable premature deaths, chronic degenerative diseases, deformations, and otherwise needlessly wasted lives.

These shattered human lives constitute but a portion of the true social costs of the acceptance, as legitimate scientific data, of the preachments of eugenics and other pseudoscientific perversions of biology and genetics by the twentieth-century lawyers, college professors, editors, jurists, legislators, governors, presidents, and voters who frame, enact, and ratify all societal policies and programs dealing with human health, human education, and human welfare.

NOTES AND REFERENCES

(1) Lidbetter, E.L., Pedigrees of Pauper Stocks, pp. 391-397, "Eugenics, Genetics and the Family," V.I. Proc. Second International Congress of Eugenics, N.Y. 1921 Baltimore, 1923; see also, Davenport, C.B., "Heredity in Relation to Eugenics," pp. 80 *et passim*, N.Y. 1911; Galton, F., "Hereditary Improvement," *Fraser's* Magazine, pp. 116-130, January, 1873.

(2) Laughlin, H.H. in "Analysis of the Hereditary Nature of Carrie Buck," Expert deposition in Circuit Court Proceedings, Amherst County, Virginia, in case of *Carrie Buck vs. J.H. Bell*, April, 1925.

(3) *Popular Science Monthly*, February, 1903, pp. 382-383.

(4) *Ibid.*

(5) Stiles, C.W. "Early History, In Part Esoteric, of the Hookworm (Uncinariasis) Campaign in Our Southern States," *Journal of Parasitology*, 33 (1947), 1-18; Williams, G., "The Plague Killers," N.Y. 1969.

(6) Parsons, "Trail to Light," New York, 1943, p. 305.

(7) Sydenstricker, E. in section on "The Prevalence of Pellagra: Its Possible Relation to the Rise in the Cost of Food," at the Third Annual Meeting. National Association for the Study of Pellagra, held at Columbia, South Carolina, October 21-22, 1915. Proc. published in pp. 2028-29, *J.A.M.A.*, Dec. 4, 1915.

(8) Goldberger, J. "The Etiology of Pellagra. The Significance of Certain Epidemiological Observations with Respect Thereto," *Public Health Reports*, Vol. 29, No. 26 (June 26, 1914), pp. 1683-86. Also: "The Cause and Prevention of Pellagra," *Public Health Reports*, Vol. 29, No. 37 (September 11, 1914); also "The Prevention of Pellagra: A Test of Diet Among Institutional Inmates," Proc. National Assn. for the Prevention of Pellagra, *J.A.M.A.*, Dec. 11, 1915, pp. 2115-2116. See also: "Goldberger on Pellagra," edited and with introduction by Milton Terris, Louisiana State University Press, 1964.

(9) Sebrell, W.H. "Clinical Nutrition in the United States." *American Journal of Public Health*, 58:11. 1968, p. 2035.

(10) Davenport, C.B. "The Heredity Factor in Pellagra." Muncey, E.B., "A Study of the Heredity of Pellagra in Spartanburg County, South Carolina." Both in *The Archives of Internal Medicine*, 18:1, July, 1916, pp. 4-75. Both also reprinted as *Eugenics Record Office Bulletin*, No. 16, July, 1916, N.Y.

(11) "Pellagra III, Third (Final) Report of the Robt. M. Thompson Pellagra Commission of the N.Y. Post Graduate Medical School and Hospital." New York, 1917. pp. 226-227, footnote.

(12) Stimson, A.M. "A Brief History of Bacteriological Investigations of the United States Public Health Service." *Supplement No. 141, Public Health Reports*, Washington, D.C., 1938.

(13) Henry S. Pritchett in, Flexner, A. "Medical Education in the United States and Canada," Bulletin No. 4, The Carnegie Foundation for the Advancement of Teaching, N.Y., 1910, p. x.

(14) Dobbing, J. in, "Nutrition, the Nervous System and Behavior," Washington, D.C., Pan American Health Organization, 1972, (Scientific Publication No. 251), pp. 21-22.

SCIENTIFIC FREEDOM: SACRED PRINCIPLE OR SECULAR POLITICS?

Ernest Drucker
and Victor W. Sidel

The controversy which today rages around the publications and lectures of Jensen(1), Herrnstein(2), and Shockley(3) may be interpreted in a number of different ways. Some aspects of the dispute are clearly "scientific" issues, a continuation of long-standing disagreements between scientists on the facts and on the scientific models and theories for their analysis.

For example: Which contributes more to the functional level of the mature adult—heredity or environment ("nature" or "nurture")? Can the measurement known as "I.Q." be determined (and has it been determined) by methods which make it free of the learned culture of the individuals tested or is it "culture-bound" and differences in it more reflective of differences in environmental influences on groups than of individual differences? To what extent do genetic differences, such as color of skin or facial characteristics, serve as "markers" or "predictors" of other possible genetic differences—such as capacity for specific kinds of learning?

These are questions which, on the face of it, seem amenable to investigation by the usual methods of science. Yet, instead of attempts to resolve the differences in the facts and in their analysis in commonly accepted ways, we see a bitter controversy which has deeply divided the scientific community.

In fact for a century or more, the framing of "heredity versus environment" questions, many of the methods of investigating this supposed dichotomy, and the quality of analysis and interpretation of findings in this area have left much to be desired. Several other chapters in this book review this history in some detail. They indicate clearly that serious logical and methodological problems are, and always have been, inherent in much of this work. Furthermore, if we accept the evaluation of many of the most prominent workers in the field(4,5), this is especially true of the body of research currently being discussed by Jensen, Herrnstein, and Shockley on racial differences in I.Q. If this is the predominant view of the quality of scientific work in the area what then could possibly be seen as so significant in the current controversy, this most recent skirmish in a long war? Does not the reaction seem far out of proportion to the importance and quality of the "scientific" issues at stake?

Closer examination reveals clearly that the vehemence lies not in the debate on the scientific issues per se, but on the potential impact of this research on social policy. Those affected either directly or indirectly by the possible applications of the research have questioned its meaning and tried to call attention to the possible harm these applications (and therefore the research) might produce—*independent of its "scientific" validity.* While public attention in the current heredity/I.Q. controversy has focused on the more militant and active expression of this protest and on the *content* of the research itself, this is only part of the picture. "Responsible" and thoughtful criticism and concern has also focused on the *manner* in which such research is conducted, how the *findings are disseminated,* and on the rights and responsibilities of the various parties—both within and outside the established scientific community. This criticism has by and large reflected the feelings of those who see real dangers in this entire line of research and feel they have the right to a say in its conduct and dissemination. Yet in many instances, both the moderate and militant calls for restraint have been rejected because they are seen to infringe on "scientific freedom." Thus many scientists react to this controversy not in terms of the content of the issue and its potential for social destructiveness (indeed many agree with the protestors), but rather see the "really important" issue as the assault on them and their freedom to perform and publicize their research.

More broadly, the controversy, both moderate and militant, occurs because the social policies in question touch on not one, but two of the most controversial issues of our time. One is that of "race," more specifically the ways in which black people have been treated by American society for 200 years and the ways in which they continue to be treated; others in this book, particularly Gerard Piel, have dealt with this issue and we will not confront it further, although we do not mean in any way to designate its importance by focusing on the other issue.

This issue, stated another way, is that this controversy touches a very sensitive nerve—i.e., the personal freedom and accountability of scientists. The purpose of this paper is to examine this view of the controversy in some detail and to question some of the basic assumptions which underlie the application of the principle of "scientific freedom" in today's world.

Considerable effort has been expended to portray the heredity/IQ issue as an example of the conflict between "scientific freedom" and political control of science by the state. However in this instance the state and its authorities are not the enemies of the scientists being attacked. Quite the contrary, the state and its authorities support much of the research in question and use it for their own purposes. The bulk of the protest in this particular instance, in fact, comes from a part of society which is relative to the scientific community dispossessed and powerless. In this case we believe a very different phenomenon is at work and the attempts to liken it to classic conflicts between "science and authority" are misleading.

Some historical perspective is of value. Since the origins of the human activity that is called "science" there has been a debate about the relation of those who practice and teach it—"scientists"—to the dominant social order and authority. Whether this authority was political, economic, religious, or magical the issue has usually been drawn in terms of two polar positions. On the one hand were those who felt the social good best served if the scientist were to work only in spheres approved by those in power, with careful control of his work and its goodness of fit with current dogma. On the other hand were those who felt that the social good would best be served if the scientist were viewed as one whose only commitment is to some objective demonstrable "truth", with not only the right but the obligation to follow that truth in whatever path it might lead no matter how painful or disturbing the consequences.

The names Bruno and Galileo in astronomy and Darwin and Lysenko in genetics bring to mind—albeit in quite different ways—these polar conflicts between the "truth-seeking" of the scientist and the "authority" of those in power. But these are only the tip of the iceberg. There have been numerous instances in which dissenting scientists have insisted on performing research or stating views which were not countenanced by the established social order. For this they were ostracized, prohibited from doing research, dismissed from their positions, or worse.

The scientist has understandably seen this pressure as a threat to his work—not to speak of his personal security—and has responded by asserting that the scientist must remain free of such restraints, if he or she is to do his best work. A special kind of "freedom"—"scientific freedom"—was defined as the right to do research and to report the findings even when the society or its representatives demanded that it not be done. From the point of view of the scientist the attempt to limit scientific inquiry is an infringement of a personal freedom—to pursue and communicate truth however he or she perceives it and wherever it leads. The "pure form" of this argument states that the principal obligation of science is this pursuit and that in itself is sufficient reason for performing scientific work. Furthermore, this decision should be made without regard to whether that work has any social consequences and, if it does have, what those social consequences are.

Obviously most scientists today would disavow this form of the argument. Its most usual form today is not that the scientist has no responsibility in the performance of research but rather a debate on whom he or she is accountable to in the assessment of that responsibility. Thus, workers such as Jensen, Herrnstein, and Shockley may claim that they are indeed performing socially responsible and socially useful work—that is, if their theories are followed social resources will not be wasted in educating the uneducable and will not lead to the frustration of the certain sub-groups trying to learn that which they are genetically ill-equipped to learn. They would argue that because of the unwillingness to look at their evidence on allegedly inherited differences in learning ability of various sub-groups, that resources which are already scarce are irrationally allocated and could be more effectively used.

The "pure form" of the scientific-freedom argument implies that scientific work lies outside the particular and changing values of states, cultures, religions or other social groups—i.e., that science is value-free or at least neutral. From this perspective the scientist, in order to do his or her best work, should be subject to no social control at all. The other form of the argument recognizes that there are value judgments involved in choosing to do certain kinds of work and in the methods used, and recognizes that these value judgments indeed imply social involvement and accountability. From this perspective the question becomes: How should that accountability be manifested?

One of our contentions in this paper is that although this second point of view is increasingly the framework for dialogue about social responsibility and science, many of the assumptions of the "pure scientific freedom" position remain as a bedrock article of faith among most, if not all, western scientists. Even if they formally deny that they believe this to be true, nonetheless scientists often act as though it were—especially in response to public criticism or political assault on the latitude of scientific work. It is particularly in those instances of conflict between science and some segment of society that we should, as scientists, most carefully scrutinize the position of science. If, as is the case in the current controversy, science invokes "scientific freedom" in its own defense—indeed as the keystone of its argument—then it is our obligation to examine thoroughly the assumptions, data, and logic of this crucial concept and its applicability to the case in question. The assumptions under which scientists operate ("scientific freedom" among them) are critical for all of us because of the enormous power and influence that technology and science holds in the modern world. When any discipline or group in the society hold such power and has a potential for such influence inherent in their activity, one must carefully scrutinize its ground-rules, its basic concepts, and its rights and responsibilities.

Since the scientific revolution (i.e., the "success" of science) there has been a strong tendency on the part of society to view these ground rules of science and technology as values and truths having a higher status than other truths and values. This is understandable in view of the enormous impact that science has had on modern life and the tendency for the non-scientist to idealize scientific work. However, scientists have played their own role in furthering this view of science. It has always been in the scientists' interest to be perceived thus, and the scientist can only too well recount periods of history when this was not the case. Thus the scientist would naturally be inclined to believe his or her own public relations and utilize prior successes to the maximum in order to garner public support and develop the power and influence of the scientific community.

This is not to say that scientists have any more prejudices, short-sightedness or tendency to advocate for their special interest than any other group. But every group guards its privileges jealously and it is our contention that scientists have no fewer of these human vices than the rest of humanity. The attempt therefore to support scientists' interests and activities by referring to higher principles may be misleading and serve to confuse matters. We take issue with the notion that the assumptions and ground rules of scientific activity are absolute or hold any higher priority than do other social concepts. In this regard the use of the concept of scientific freedom to shield scientists from scrunity by the public and accountability to the public, *especially that public affected by the consequences of their work or of its dissemination*, is suspect. It is our judgment that the more common applications of this concept represent an abuse, a form of special pleading or political activity rather than a statement of higher principle.

Another important element of our position relates to the composition and characteristics of the scientific community itself. The manner in which scientists are selected, the training and patterns of support via which scientists and scientific work have developed, and the role and position of the scientific community and modern technology in the social order strongly suggest that the abstract and absolute concept of scientific freedom, as it has been used historically, has lost much of its meaning.

While it has always been true that science and scientists are in no special way "objective" or "neutral" and in no sense free of the influence and pressures of the society in which they live, the particular pressures on scientists in modern society are even more pronounced than in the past.(6) These pressures determine who will be a scientist, determine his or her career choice, define professional success, and powerfully affect the areas in which "science" is done. In fact science today is under enormous "social control." This can be seen by examining just a few of the factors which select and mold scientific personnel and their activities.

Selection: The population from which scientists are chosen is by no means representative of the population as a whole. This selection process is defined and enforced by a community of scientists in authority who serve the function of selecting scientific abilities in and for a particular society. The socio-economic level (that is, the social class) of a student often determines whether talents will be nurtured, and in what direction. In general scientists are chosen from among the elite in the society.

Training: Scientific training is synonymous with a particular style of education. Its application to research (if not quite dogma) is a rather rigid orthodoxy. Deviations are permitted only within closely prescribed guidelines. If one wants to "get along" one had better "go along" (e.g., in the choice of areas and methods for dissertation research in the university).

Role Models: Most role models in science reinforce conformity with prevailing institutional patterns and a clannish isolation from the larger society. Even where the role model represents deviations from conformity it is an elitist model: the scientist who professes to follow "his or her own conscience or insights" often forgets that conscience and insight have societal roots.

Career Development: The professional career of the scientist is fraught with pressures toward conformity and toward work in areas prescribed by the interest of senior researchers, the strengths of university departments, or the financing arrangements of particular government agencies. The status system of the scientific profession is organized on a hierarchical basis with the young scientist expected to follow an orderly progression up a ladder of increasing prestige, privilege and power.

Support: Except in very exceptional circumstances scientists must work on projects for which funding exists. The selection of topics for research, methods for research, and methods of communication of results often depend on the policies of such funding sources.

Communication: The dissemination of scientific ideas and data is usually couched in terms, often jargon, which will appeal to other scientists and to funding agencies. Rarely is it directed to the world outside the scientific establishment. It reflects a set of values of what is important and who is important.

Put another way, the scientist in modern society is party to what Spencer Klaw has called a "Faustian bargain"(7), the full import of which he may not appreciate until he has gotten deeply into his professional career. Individuals who become scientists are selected, trained and come to function in ways which generally (rather than exceptionally) confine and constrain the areas of investigation, the style of investigative approach and, in some cases, the findings themselves.(8,9,10) In recent years it has become clear that scientists for the most part have been permitted fields of investigation which have been defined by a scientific establishment and selectively encouraged or discouraged via the mechanism of research financing.

In short, the notion that scientists operate in a place where the everyday pressures of power, prestige, and even profit are not operative is a myth. What the scientist does is shaped by societal forces, to a degree at least equal to those influencing the non-scientist. Indeed his or her work may be seen as largely a product of these forces. The question addressed by this paper may therefore now be recast not as *whether* there *should be* social control of science and the scientist but rather *by whom* (and therefore for whose benefit) that control is exerted. This question becomes more urgent as the consequences of scientific activity on the environment, on the social order and on the perception of the psychological nature of man himself become more influential and therefore more potentially destructive.

What, then are the alternatives available for this social accountability?

The first is to continue the status quo, with scientists either deceiving themselves or deceiving others with their pretense of "scientific objectivity" or "neutrality". This course—which is often defended under the banner of "scientific freedom"—in our view represents neither freedom, neutrality nor objectivity.

The second course, in many ways simply a more socially-acceptable modification of the first, has been an attempt at "self-regulation" by scientists. This has usually taken the form of sounding warnings about—and at times self-regulated abstinence from—certain lines of research which seem particularly socially destructive. Scientists have called upon their fellow scientists to voluntarily refrain, at least temporarily, from specific lines of research until the risks that might be inherent in them might be assessed. In other instances they have themselves, as individuals, refused to participate in research which they considered potentially destructive; post-World-War-II examples include that of the physicists who questioned the wisdom of further development of even more destructive nuclear weapons(11) and of those engaged in biological and chemical warfare research, such as Theodor Rosebury, who raised issues of ecological and social destruction.(12) More recently eleven distinguished American molecular biologists published a letter in the two most prestigious U.S. and British general scientific journals, *Science* and *Nature*, urging their colleagues to halt two types of gene-transplantation experiments until "attempts have been made to evaluate the hazards and some resolution of the outstanding hazards has been achieved."(13) This attempt at self-regulation seems to go beyond the others since it is concerned with "pure" or "basic" research rather than "applied" research on weapons or other forms of technology.

In some ways these attempts at self-regulation are indeed laudable, but as Tabitha M. Powledge recently phrased it:

> That eminent scientists choose to say they are doing something possibly dangerous and that they want to stop for a while to think about it ought to be cause for dancing in the streets. But it would certainly be foolhardy to rely routinely on the goodwill and good sense of eminent scientists—and even more foolhardy to rely on those who aren't so eminent but want to be.(14)

The third course is some sort of direct regulation by non-scientists. But this course, too, is fraught with difficulty, as experiments with "community control" of schools, consumer control in health care(15), and the attempts to involve the public in other technological institutions have demonstrated. Most laymen in the U.S. (and even many scientists when out of their own narrow fields) are abysmally ignorant of the content and techniques of science and feel intimidated by highly specialized jargon. Furthermore, community members (especially members of poor communities) are generally unwilling to take the time to become knowledgeable in order to make considered decisions of scientific issues. Perhaps they are conditioned by a long history of having no effective method for bringing any significant influence to bear on social institutions—particularly the technology-based ones.

Given these difficulties and the social and political realities of this society it is unlikely (perhaps even undesirable) that direct regulation or community control of science will occur in the near future. The immediately available alternatives seem unlikely to be successful in broadening the base of accountability of science, although the problems of increasingly concentrated power and influence by a narrow range of social forces over science and of increasing power and influence of science over social policy will persist. Where may we turn? Our partial answer is that we must explore methods which convert the issue from one of control of science to one of reducing the distance between scientists and the people their work is meant to benefit. In the long-run these methods should reduce, and in some ways eliminate, the distinction between scientist and non-scientist.

In the meantime, the scientist should accept the responsibility for *adequate communication.* The scientist has the obligation to make clear the nature of the subject of his or her research plans and the methods he or she has used or will use to those it might affect. The scientist must state the assumptions underlying the work as specifically as he or she can and the justification for doing this work at all relative to other possible uses of time and resources. Most important, the scientist must consider and communicate as clearly as he or she can the possible consequences of the work in terms of the social action, changes in technology, and the further research which may be based upon it. This communication must be presented in language that non-scientists can understand so that those outside of the scientific establishment can become involved in the process of allocation of scarce resources and other priority decisions.

To those who respond that it is often impossible for a scientist—particularly for one working in basic science—to visualize the potential consequences of his or her work we say that the attempt must be made. Scientists, even those engaged in basic research, are usually quick to perceive and explain the "relevance" and "importance" of their work when seeking public funding for it; the same ability should be available to the scientist when the issue is not public funding but public accountability. The scientist's technical knowledge of a subject should enable him or her to predict at least some of the possible consequences of the work. If the scientist cannot foresee all the consequences the very act of attempting to make such evaluations may cause him or her to alter the direction or methodology of the work—or to abandon it entirely. More important it will give those who make decisions on funding, publishing or otherwise supporting the work some tools with which to make their decisions in ways which—however imperfectly and with great distance from those affected—improve the consideration of the social impact of the work.

A second responsibility is the *obtaining of informed consent* of those from whom or on whom the data are gathered. The scientist who deals with inanimate or non-human subjects or long-dead societies of course cannot get permission from his subjects. Here we speak rather of a responsibility peculiar to the student of current society or to the clinical scientist. That such informed consent must be obtained from the specific individuals who, via sampling or other selection participate in the research, is now well established, even if at times honored more in the breach than in the observance. There has in recent years, for example, been increasing pressure on institutions in which clinical research is done to set up committees to approve and monitor research on human subjects and to make certain that appropriate legally effective informed consent is obtained. But we advocate going much further than that. Where identifiable sub-groups in the population are defined (in the sense that the data gathered may be generalized to all members of that sub-group) informed consent must be obtained from representatives of the sub-group as well. Thus those who wish to study the Navaho culture, for example, must seek the consent not only of the specific people whom they interview but of the representatives of the Navaho people as well. This responsibility must extend to those who analyze others' data if the data are to be used to reach conclusions about current issues or still-existing groups. Since researchers dealing with secondary material will very rarely have access to original respondents, the informed consent of the representatives of the affected sub-group is all the more important.

Obviously there can and will be a range of reactions with any large group and a "representative" opinion is usually difficult to define. But the acceptance of the responsibility of *soliciting* that reaction, the need to justify the work to those affected, and, in the absence of concensus, to permit a third party or agency to arbitrate differences, seem to us wholly different and inherently more just approaches than the one currently used—i.e., the lone discretion of the scientist and his or her institutional representatives. If the objection is made that this will severely limit the kinds of social and clinical research which can be done, we say that the establishment and observance of such a process would do more to improve the quality of life in our society than the possible findings of the research that it might inhibit. In our view there can be no right to do research until it is granted by those from whom the data come and to whom the "data" belong.

A third basic responsibility of the scientist is *vigorous advocacy of the rights of those affected* by the work and its consequences. This responsibility, by far the most difficult, is much broader than the second and encompasses social and natural scientists alike. To say that consequences at times cannot be foreseen and that those affected therefore cannot be identified does not change the responsibility of those doing research where consequences can be foreseen, however dimly. This responsibility is so strong, it seems to us, that the scientist must lean over far backward to make certain it is met. This principle has already been adopted by the ethics committee of the American Anthropological Association—it states that the first obligation of the researcher is to his informants, i.e., those he or she has studied.

We do not contend that these proposals would be easy to implement or that these are the only proposals which would meet the goal of a more broadly responsive and accountable science. Other societies, such as China, are experimenting with other, more far-reaching methods.(16) Nor do we contend that these proposals will not lead to other problems—problems which in some ways will be as troublesome as those they are designed to meet. Our contention rather is that the larger problem, of which the current heredity and intelligence controversy is merely a symptom, is a very critical one, that science and scientists are shaped by a very narrow and unresponsive set of social forces, and that science and scientists have reached positions of very great and often uncontrolled power in the society. We believe a huge gulf exists in our society between scientists and those whom their work will affect in increasingly profound ways. Until ways are found to bridge this gap we will find ourselves in ever deeper trouble.

NOTES

(1) Arthur Jensen. "How much can we Boost IQ and Scholastic Achievement?" *Harvard Educational Review*, Vol. 39, No. 1, 1969.

(2) Richard Herrnstein. "IQ." *Atlantic Monthly*, 228, September, 1971.

(3) William Shockley, "The Apple-of-God's-Eye Obsession." *The Humanist*, Jan.-Feb., 1972.

(4) Walter Bodmer and Luigi Luca Cavalli-Sforza. "Intelligence and Race." *Scientific American*, Vol. 233, No. 4, October, 1970.

(5) Jerry Hirsch. "Jensenism: The Bankruptcy of 'Science' Without Scholarship." *Educational Theory*, Vol. 25, No. 1, Winter 1975.

(6) Max Black. "Is Scientific Naturality a Myth?" Lecture delivered at the Annual Meeting of the American Association for the Advancement of Science, New York, January 27, 1975.

(7) Spencer Klaw. "The Faustian Bargain." *The Social Responsibility of the Scientist*. Edited by Martin Brown, New York: The Free Press, 1971, pp. 3-15.

(8) Jane E. Brody. "Charge of False Research Data Stirs Cancer Scientists at Sloan-Kettering." New York *Times*, April 18, 1974.

(9) Barbara J. Culliton. "The Sloan-Kettering Affair: A Story Without a Hero." *Science*, Vol. 184, May 10, 1974, pp. 644-650.

(10) Ernest Borek. "Cheating in Science." New York *Times*. January 21, 1975.

(11) Robert Gilpin. *American Scientists and Nuclear Weapons Policy*. Princeton: Princeton University Press, 1962.

(12) Theodor Rosebury. *Peace or Pestilence: Biological Warfare and How to Avoid It*. New York: Whittlesey House, 1949.

(13) Paul Berg, et. al. Letter to the Editor: "Potential Biohazards of Recombinant DNA Molecules," *Science*. Vol. 185, July 26, 1974, p. 303.

(14) Tabitha M. Powledge. "Dangerous Research and Public Obligation," New York *Times*, August 24, 1974.

(15) Victor W. Sidel. "Quality for Whom? Effects of Professional Responsibility for Quality of Heath Care on Equity," *Bulletin of the New York Academy of Medicine* (in press).

(16) *China: Science Walks on Two Legs*. A Report From Science For the People. New York: Avon Books. 1974.

"...YE MAY BE MISTAKEN"

Gerard Piel

I beeseech ye, in the bowels of Christ, think it possible ye may be mistaken.
—Oliver Cromwell to the General Assembly
of the Kirk of Scotland, August 3, 1650
(Carlyle, 1845)

In a statement under the title, "Comment" fifty scientists have brought serious complaints against colleagues and others un-named. They charge the professional offense of orthodoxy to "academics committed to environmentalism in their explanation of almost all human differences." They allege that those who "express a hereditarian view" of human behavior have suffered "suppression, censure, punishment and defamation" by scientists and "nonscientists and even antiscientists." (*American Psychologist*, 1972)

Of this statement it must first be observed that it belongs to the category logicians call "autological." That is: it is true of itself. The complaint of breach of the peace is itself a breach of the peace. The charge of orthodoxy asserts an orthodoxy. In sum, the rules of civil discourse that govern the polity of science have been suspended by at least one if not all parties to this controversy.

Controversy is not strange to science. As the work of science may be more confidently described as the correction of error than the discovery of truth, so the adversary process is its life. Discourse is ordinarily addressed to the substance of statements, to the integrity of the evidence and the rationality of its interpretation. All parties are bound to "the habit of truth" (Bronowski,

1956) by the certainty of exposure if they depart from it.

The present controversy has escaped these bounds. This is one of those occasions, epitomized by Robert Merton, when the parties "no longer inquire into the content of beliefs and assertions to determine whether they are valid or not, no longer confront the assertions with relevant evidence, but introduce an entirely new question: how does it happen that these views are maintained?" (Merton, 1973) It is to the credit of behavioral scientists that such occasions do not arise more often in their deliberations. Their work is beset by a double uncertainty. The act of observation perturbs not only the event but the observer; the study of human behavior ineluctably engages human behavior. Nowadays the work comes under still another strong perturbation, from outside of science. Rationalizations supplied from the social sciences are much favored by the makers of public opinion and policy. In acknowledgement of this tripling of the uncertainty, Gunnar Myrdal has urged that social scientists, upon searching self-examination, should surface and declare their bias and make the open test of evidence against that declaration an integral step in the development and presentation of evidence. (Myrdal, 1969) This would surely prove a more fruitful as well as more honorable procedure than the conventional pretense that the conclusion has followed from innocent Baconian induction.

Thus, it clarifies the present controversy to acknowledge that it involves not only behavior and heredity but race. That explains much of the incivility it has occasioned. Americans can find some comfort in the knowledge that racism is not peculiar to their institutions and customs. To make a "pseudo-species" (Erikson, 1969) of a group of fellow men is a crime against humanity that has been committed by virtually every known human society, except possibly the lonely Copper Eskimos. In contemporary history the record shows no exception. The Pakistanis and the Indians; the Pakistanis and the Indians again, this time together as refugees from East Africa and victims of discriminatory immigration barriers erected by a British Labour Government; the mutual slaughter of Ibos and Yorubas; Jews and Arabs; Catholics and Protestants in Ulster; Turks and Greeks in Cyprus—the roll-call could go on quite indefinitely, summoning victims of ancient folk prejudice and innovative state policy from all around the world. As is evident, race is not essential to this social pathology. Race and racism serve, however, to exacerbate it. Our country's affliction is complicated still further by the fact that our principal racial minority was held in chattel slavery for 200 years.

As Barbara Jordan observed in the course of the recent deliberations of the House Committee on the Judiciary, "We the people . . ." did not include the likes of her until, at the earliest, 1868. The Reconstruction brought *de jure* but not *de facto* change in the status of the American Negro. By 1875, under the so-called civil rights decisions of the U.S. Supreme Court in that year, the legal gains had been largely nullified, especially in the South. The Second Reconstruction may be said to have begun with the *Brown* decision in 1954, at the extreme reach of the already ebbing tide of benevolence that had got under way in American politics in the 1930s. For behavioral scientists that decision had special meaning. It cited the work of behavioral scientists in affirmation of the ethical ideals of American society, and it placed the weight of the Constitution behind ameliorative measures to enhance the behavioral development of the country's youth.

Observance and enforcement of the law of the land came uncertainly against cross-currents of reaction and backlash. As late as 1963, John Minor Wisdom had to write for Fifth Circuit Court of Appeals in New Orleans:

> We again take judicial notice that the State of Mississippi has a steel-hard, inflexible, undeviating policy of segregation. The policy is stated in its laws. It is rooted in custom. The segregation signs at the terminals in Jackson carry out that policy. The Jackson police add muscle, bone and sinew to the signs. (*U.S. v. City of Jackson, Miss.* 1963)

This time the hardening of racial lines in the South signified the technological obsolescence of the black field hand; the postwar migration of the American Negro was carrying the heritage of 12 generations of political and social oppression from the rural slums of the South into the ghettoes of the country's major cities. An obdurate new pattern of metropolitan segregation (Grodzins, 1957) now divided American society in every major population center. The ugly reality of racism is measured by such indexes as: infant mortality that runs at twice the rate among black as among white Americans; unemployment, three and four times as high among young black males; populations in prisons, houses of detention, mental hospitals and children's shelters that are 85 percent black; a differential of 2.6 and 2.0 years in schooling as between whites and blacks, male and female, respectively; a three-times overrepresentation of blacks in the income groups below the poverty line. The degradation of caste is enforced not only by "muscle, bone and sinew" but also by benevolence—as by the concerting of welfare measures that pay the least for support of needful children at home and the most to institutionalize them. (Polier, 1973)

Under the moral leadership of a national administration too squalid to stay in office, Americans in every major city have now risen, on the fraudulent issue of "busing," to place the power of public funding behind racial segregation in the country's schools and neighborhoods. In the very hour of that administration's fall, the U.S. Supreme Court that bears its mark, in a 5 to 4 decision on the busing issue, (*Milliken v. Bradley*, 1974) effectively frustrated the *Brown* decision. The South, having yielded its *de jure* segregation is now free to import *de facto* racism from the North.

It was to this crisis of the nation's soul, to the waning of the Second Reconstruction, that certain bio-behaviorists contributed the papers, the articles and the public appearances that are at issue in the present breach of the polity of the behavioral sciences. I have here reviewed the historical background in order to explain, in part, the passions they aroused. The principals claim they have been misquoted and misrepresented. Because my enquiry into this episode requires consideration of the primary documents, I must now expose myself to that charge.

The leading contribution, of course, was that of Arthur Jensen, published in the Winter 1969 issue of the *Harvard Educational Review.* (Jensen, 1969) In that journal, the paper served as a survey of work and of questions largely outside the competence of its editors and readers. It is, in fact, a secondary review of a large body of literature, with little primary reference to original work by the author. Up to a point it stands as a creditable popularization of the principles of population genetics and a review of efforts to estimate the relative contributions of heredity and environment to the shaping of the behavioral competence called "intelligence."

Jensen is careful to declare that "heritability"—the unit in which estimates of the genetic contribution are expressed—is a "population statistic," having "no sensible meaning with reference to a measurement or characteristic in an individual." He insists that heritability "is not a constant like π and the speed of light." As a population statistic, its value is affected by the characteristics of the population: "It will be higher in a population in which environmental variation relative to the trait in question is small, than in a population in which there is great environmental variation ... (The) value of H is jointly a function of genetic and environmental variability in the population. Values of H ... do not represent what the heritability might be under any environmental conditions or in all populations or even in the same population at different times. Estimates of H are specific to the population sampled...."

From this point, Jensen's discussion proceeds outside the restraints these careful statements imply. Principally from a review of evidence from studies of the I.Q. of identical twins, conducted uniformly among white American and British children, he postulates .80 as the value of the heritability of intelligence, applicable to black as well as to white Americans. To make this postulate stick he must show that the contrasting environments that nurture whites and blacks in America exert the same .20 residual effect upon the expression of their genes for intelligence. The result is what accountants call a "plugged" figure. Jensen holds the black environment down to .20 by allocating genial discounts to prematurity, prenatal and perinatal nutritional deficits, maternal and even sensory deprivation in infancy, cognitive dissonance between home and school, and the segregation and tracking that give the overwhelming number of black children inadequate schooling. He does not mention another factor cited in the Coleman Report, "Equality of Educational Opportunity," as one that had a "stronger relationship to achievement than ... all the 'school factors' together." This is "the extent to which an individual feels that he has some control over his own destiny" a feeling that was found nearly non-existent in black children except in those few schools where their representation approximated that of blacks in the population. (Coleman, 1966)

In this context, Jensen considers the comparative distribution of I.Q. scores in the black and white American populations. The mean for blacks invariably falls one standard deviation below the mean for whites. The bell curve for the blacks thus places 85 percent of them to the left of the mean for whites and the majority below the 75-I.Q. "retarded" line. The intelligence of blacks is depressed by their heredity, therefore, to the point where most of them are genetically retarded.

The proposition that .80 stands as a kind of constant, making it possible to compare genetically fixed intelligence in different populations, is put forward at the outset as a hypothesis. It serves thereafter, however, as a given. The discussion that follows builds a multi-storeyed characterization of the genetic endowment of the black race on this wholly un-secured foundation.

The pivotal error in the Jensen argument has been called for what it is by Theodosius Dobzhansky and by Walter Bodmer and Luigi Luca Cavalli-Sforza. After chiding Jensen for his misappropriation of the concept of heritability, Dobzhansky goes on to dissent firmly from his discount of the depressing forces of the black ghetto environment. (Dobzhansky, 1973) Bodmer and Cavalli-Sforza conclude: "... there is no logical connection between heritabilities determined within either race and the genetic difference between them." (Bodmer and Cavalli-Sforza, 1970)

In a closer inspection the evidence from the same twin studies surveyed by Jensen, Urie Bronfenbrenner casts considerable doubt upon the value of .80 for the heritability of intelligence even in the white population. It turns out that the twins who were presumably reared "apart" were mostly raised in the same towns, in the same socioeconomic strata, in households of the same family and went to school for the same number of years in the same or similar classrooms. For these twins, I.Q. tests showed correlations indicating a heritability of intelligence above .80. For those few pairs who were raised in unrelated households, however, the coefficient was .63; where they were raised in different towns, the coefficient was .66, and for those raised in dissimilar communities (e.g.: mining vs. agricultural) the coefficient fell as low as .42 in one sample and .26 in another. In fact, from consideration of these data, the original authors of one of these studies concluded that "differences in education and social environment produce undeniable differences in intelligence." Bronfenbrenner is led to the parallel conclusion that "the heritability coefficient should be viewed not as a measure of the genetic loading underlying a particular ability or trait, but rather as an index of the capacity of a given environment to evoke and nurture the development of that ability or trait." (Bronfenbrenner, in press)

To the much same effect is the result of a study by Sandra Scarr-Salapatek that considered the I.Q. correlations in more than 1,000 sets of identical twins from different racial and social class groups. She found that "... total variance (in I.Q.) attributable to genetic sources was always higher in the advantaged groups of both races." And, contrariwise, "In most cases, genetic variance could not be estimated for the aptitude scores of the lower-class children." (Scarr-Salapatek, 1971) In other words, "talent" may manifest itself conspicuously in people who grow up in favorable environments and remain suppressed in adverse environments.

Such a finding holds no surprise for a geneticist. Jerry Hirsch, a psychiatrist turned geneticist, calls the attention to biobehaviorists to the "norm of reaction," as the measure of the protean character of the genotype in its dynamic relationship to the environment. In plants and in some animals, the range or norm of reaction has been estimated from replication of the genotype in a wide range of environmental conditions. "The more varied the conditions, the more diverse might be the phenotypes developed from any one genotype." On the other hand, "Of course, different genotypes should not be expected to have the same norm of reaction." Such observations sink without trace the environmentalist cliché: "heredity sets the limits, but environment determines the extent of the development within those limits." At the same time, the new hereditarians should learn they are mistaken to "assume an inverse relationship between heritability magnitude and improvability by training and teaching ... Heritability provides no information about norm of reaction." (Hirsch, 1970)

The Jensen paper was published without benefit of informed criticism of this kind. Unencumbered by the subtlety and power of well-demonstrated concepts from genetics, Jensen was able to proceed to the questions of public policy that interested him more. He is satisfied that the genetic deficiency of blacks explains the disappointing results reported so uniformly—starting with the Coleman Report in 1966—from studies of ameliorative and compensatory educational programs. Since the genotype defies manipulation by such well-meaning efforts, it is a waste of scarce educational resources to try to upgrade the I.Q.s of young blacks.

At about this point in his argument, Jensen was seized by a fresh concern: Is the "national I.Q." declining in consequence of the higher procreative propensities of the lower classes, especially the lowest class blacks? "No one should be more concerned about this than the Negro community itself ... Is there danger that current welfare policies, unaided by eugenic foresight, could lead to the genetic enslavement of a substantial segment of our population?"

There is the odor here of the classical hereditarian anxiety put abroad by Francis Galton among his Victorian contemporaries when Social Darwinism reigned. To this primitive proposition and the social policy it implies, Dobzhansky replies that it is "hypocrisy to say that the deprivations of the poor come from their genes and must be corrected by eugenic elimination of those genes." (Dobzhansky, 1973) Bodmer and Cavalli-Sforza, from knowledge of rates of genetic change in natural selection, find that not even chattel slavery could have exerted sufficiently rigorous social selection against genes for high I.Q. in the American Negro population. (Bodmer and Cavalli-Sforza, 1970) Sir Cyril Burt, whom Jensen cites as a supporting authority for fixing the heritability of intelligence at .80, was cheerfully confident that the gene pool of the British lower classes continues, through vertical mobility, to replenish the I.Q. endowment of the country's upper classes. (Burt, 1961)

Jensen was putting a new gloss on Social Darwinism, however, making social stratification the outcome, rather than the agent, of natural selection. The very egalitarianism of American society, by reducing barriers to genetic competence, is in process precipitating blacks and other incompetents out at the bottom. This idea was to grip the imagination of Richard Herrnstein and inspire his contribution to the present controversy.

For egalitarian nemesis, Jensen has a social-engineering prescription. Recognition of the peculiar Negro genetic endowment will show us how to arrange an appropriate place for black people in America. Studies demonstrate that black infants have a higher D.M.Q.—developmental motor quotient—than white; that is, they exhibit developmental precocity and more rapid attainment of motor coordination. What is more, it turns out, the preoccupation of educational testers with I.Q. has been obscuring a capacity for "associative learning," as distinguished from the higher-order "conceptual learning" that dominates the I.Q. scale. Citing his own work, Jensen is able to testify that black children score high on this "learning quotient" and are eminently teachable at the level of the three R's. It is their deficient capacity for conceptual thinking that depresses their I.Q. scores. He reports further that this skew in the learning abilities of blacks is shared by white children of lower socio-economic status: "The genetic factors of each of these types of ability are presumed to have become differentially distributed in the population as a function of social class." From this background, Jensen makes the almost unexceptionable recommendation that the educational system "develop techniques by which school learning can be most effectively achieved in accordance with different patterns of ability."

The only objection is to caste—to the sorting out of young people for different kinds of education and so different life roles by such genetic markers as skin color. The objection would come from those of our fellow citizens who would refuse to accept their allotted places in the brave new world that Jensen builds upon his defective genetics. In Jensen's model of their genotype, American blacks can readily recognize the admiring condescension with which slaveholders were pleased to characterize their agile, teachable chattels. In resistance to such reconstruction of the American social order, the black minority would be joined by those whites who discover they too have been similarly pigeonholed, without the dignity of divination by skin color.

The Herrnstein article, in the *Atlantic Monthly* for September 1971, (Herrnstein, 1971) is addressed to this penchant for equality that animates the political life of all peoples who aspire to self-government. As a reprise, in part, of the Jensen paper, it constitutes a tertiary treatment of questions of race, heredity and behavior, with Herrnstein's own socio-philosophical speculations supplying the principal novelty.

Herrnstein opens with a lengthy review of the evolution of the I.Q. test. This is useful because it reminds the reader how early in its history this test in children of acquired knowledge, classroom skills and simple rote learning became invested with the power of insight into the endowment of the genetic homonculus—what Ethel Tobach calls the cryptanthroparion—lurking somewhere in the child. Significantly, Herrnstein makes no mention of the work at the Child Welfare Research Station at the University of Iowa in the late 1930s that demonstrated the mutability of the I.Q. in response to favorable and unfavorable changes in the children's circumstances. (Stoddard, 1943) He cites the famous studies of M. Skodak and H.M. Skeels for their showing of correlation (in rank order) between the I.Q. of foster children and their biological mothers, but not for the 20 point average spread in the I.Q. of those children over their mothers. (Skodak and Skeels, 1949)

"The problem with nature and nurture," says Herrnstein, "is to decide which—inheritance or environment—is primary." This non-problem is the King Charles' head of the behavioral sciences. It has obscured real questions and blocked the development of a conceptual apparatus, such as that possessed by genetics, appropriate to the wealth and complexity of behavior, starting with the lowest infra-human organism capable of this sign of life. T.C. Schneirla once proposed that the dichotomy be restated in the semantically neutral terms "maturation" and "experience"—" 'maturation' . . . to refer to the contributions to development from growth and tissue differentiation, together with organic and functional trace effects persisting from earlier development" and " 'experience' as the contributions to development of the effects of stimulation from all sources (internal and external), including their functional and trace effects surviving from earlier development." In consequence of feedback linkages between internal and external events and between earlier and later stages of development, "The developmental contributions of the two complexes, maturation and experience, must be viewed as fused (i.e.: inseparably coalesced) at all stages in the ontogenesis of any organism." (Schneirla, 1966) This is a prescription, of course, for hard work. Furthermore, it underlines how little work has been done on the longitudinal development of behavior in any organism and especially in man.

The interest that Herrnstein's treatment of his non-problem holds is largely external to his presentation. In effect, he rehearses at greater length the Jensen proposition that pursuit of equity in our social order must result in its biostratification, with those of lowest genetic endowment, especially the blacks, precipitated out at the bottom. Of greater significance is his vision of who would be selected up to the top. With frictionless vertical mobility, the meritocracy would come under the dominance, of course, of its highest I.Q.s. From his lofty vantage, at six times the income and the status of fellow citizens in the bottom socioeconomic stratum, the professor forgets that there tower above him strata of income, prestige and power reaching up to six times and more his own. This, if not a decent humility in the possession of inherited endowment, should dampen his self-adulation. A behavioral scientist ought to recognize, moreover, that the rough and tumble of getting and spending and of getting the business of society done selects for other characteristics as well as intelligence. Dobzhansky lists "persistence, willingness to work, originality, creativity, leadership, ability to get along well with other people and plain human decency" and suggests these may be grounded in genetic endowment. (Dobzhansky, 1973) History shows that still other qualities count in the management of our society and may be heritable. Thus, the generalship at Attica in 1971 of the then governor of New York brings to mind the mettle displayed by his grandfather in the painful events at Ludlow, Colorado, in 1914. (U.S. Industrial Relations Commission, 1916)

It is this present, imperfect society that Herrnstein, in the end, urges upon his black fellow citizens. They should abandon their agitation for equality lest they win the self-defeating victory that would expose the poverty of their genetic endowment. In sum, they should stop reminding us of their accusing presence.

Of the social psychology and population genetics of William Shockley, there is still less to be said. One example of his imperfect ratiocination will suffice. "My own preliminary research," he writes, "suggests that an increase of 1 percent in Caucasian ancestry raises Negro I.Q. about one point on the average for low I.Q. populations." (Shockley, 1971) From Bodmer and Cavalli-Sforza this drew the comment that Shockley's Caucasian gene must have such potency as to "make a black with 50 percent white genes more intelligent on the average than an average white, and a black with 100 percent white genes (who would be an average white) an absolute genius." (Bodmer and Cavalli-Sforza, 1971)

As neither a hereditarian nor an environmentalist, but as a reader and as an observer of the events in consideration, I can assure these three authors that it is not their science but their moral philosophy that has stirred the controversy of which they complain. Their biology has flunked the examination of every qualified geneticist who has put himself to the trouble. For their sociology and political science, the precis I have given you will have to suffice. This insubstantial stuff has claimed attention in the polity of science, I conclude, only because these authors and their sympathizers have attracted such attention outside.

Out there on the hustings, they are entitled to the immunities of citizenship. But they may not honorably claim the sanctuary of science for any of the material it has been my unpleasant duty to review here. None of it was offered as primary work for the consideration of colleagues competent in their fields or in the fields into which they blundered. In each case, they made their offering outside the polity of science to audiences that were bound to be mistakenly impressed by the authors' purported scientific authority on the questions they addressed.

Shockley presents a special case. Here we behold a professor engaged in the unbecoming conduct of provoking troubled and reactive undergraduates. (Singer, 1974)

To the extent that the experience of these three authors underlies any of the charges made in the "Comment" on heredity and behavior, I believe that a great many of the 50 signators were either uninformed or misled. In particular, I can assure all signators that the espousal of "the hereditarian view" is not the cause of the troubles in question here. Jensen, Herrnstein and Shockley have joined a much larger group of social scientists in the disdain of the same agitated groups of scientists, non-scientists and antiscientists. The names of Edward C. Banfield, (1970) James S. Coleman, (1966) Christopher Jencks (1972) and Daniel Patrick Moynihan (1972) lead this supplementary list, and none of them are practicing hereditarians. What identifies all of these scientists, hereditarians and non-hereditarians alike, is that they have loaned their authority to public policies and popular attitudes associated with the braking and reversal of the Second Reconstruction—and so to the ominous division of our society on racial lines.

From my own reading of these other authors, I can testify that not all of their work is so lacking in merit as that which I have reviewed here, although I am compelled to regard some of it as more so, and more mischievous as well. Some of this work has been misinterpreted in transmission by the unstable amplifier of our popular press; some of it has been misused by public officials. It is plain that just about all of it would have been ignored, if it had not served the purpose of particular economic and political interests. The scientists honored by such recognition should remember that the incorporation of their work in public policy does not constitute its validation, powerful as events may prove the principle of the self-fulfilling prophecy.

I hope these scientists will watch closely the uses to which their work is put and that they will be alert to the human consequences. They may then come to understand the springs of the disorder and violence so priggishly deplored in "Comment." That is but a leak around the furnace door of the contained violence that imposes second-class citizenship in Jackson, Mississippi; in Watts; in South Chicago; in Detroit; in Harlem, and now even in Coney Island. If as scientists they have regard for the truth and abhorrence of error, they must as citizens tremble before the awful responsibility to which they have presumed.

I have here set down evidence of defect in the intellection of these scientists. Events have shown they have no heart. Distaste for the poor is a common and mean enough middle-class sentiment. (Edmond, 1973) These workers in the uncertain realm of the behavioral sciences, armed with a jejune nature-nurture dichotomy, have done injury to children. They have helped to abort the ameliorative and compensatory pre-school and elementary school programs our society has surely long since owed on much larger scale to the children of the poor and black.

REFERENCES

American Psychologist, July, 1972.

Banfield, Edward C. The Unheavenly City. New York: Little Brown, 1970.

Bodmer, Walter and Luigi Luca Cavalli-Sforza, "Intelligence and Race," *Scientific American*, Vol. 223, No. 4, October, 1970. Letters to the Editor, *Scientific American*, Vol. 224, No. 1, January, 1971.

Bronfenbrenner, Urie, "Nature with Nurture: A Reinterpretation of the Evidence." In Race and I.Q. M.F.A. Montague, Editor, New York: Oxford, In Press.

Bronowski, J. "Science and Human Values," *The Nation*, 27 December 1956 and "The Abacus and the Rose." New York: Harper & Row, 1965.

Burt, Cyril. "Intelligence and Social Mobility," *British Journal of Statistical Psychology*, 1951, 14, 3-24.

Carlyle, Thomas. Oliver Cromwell's Letters and Speeches with Elucidation, Vol. II. New York: Wiley & Putnam, 1845.

Coleman, James S. et al. Equality of Educational Opportunity. Washington: U.S. Government Printing Office, 1966.

Dobzhansky, Theodosius. Genetic Diversity and Human Equality. New York: Basic Books, 1973.

Edmond, Ron et al. "A Black Response to Christopher Jencks' Inequality and Certain Other Issues," *Harvard Educational Review*, Vol. 43, No. 1, 1973.

Erikson, Erik. Gandhi's Truth. New York: Norton, 1969.

Grodzins, Morton. "Metropolitan Segregation," *Scientific American*, Vol. 197, No. 4, October, 1957.

Herrnstein, Richard. "I.Q.," *Atlantic Monthly*, 228, September, 1971.

Hirsch, Jerry. "Behavior-Genetic Analysis and the Biosocial Consequences," *Seminars in Psychiatry*, Vol. 2, No. 1, February, 1970.

Jencks, Christopher. Inequality. New York: Basic Books, 1972.

Jensen, Arthur, "How Much Can We Boost I.Q. and Scholastic Achievement?" *Harvard Educational Review*, Vol. 39, No. 1, 1969.

Merton, Robert. "Paradigm for the Sociology of Knowledge." In: The Sociology of Science. Chicago: University of Chicago Press, 1973.

Milliken v. Bradley, 418 U.S. 717 (decided 25 July 1974).

Moynihan, Daniel Patrick and Frederick Mosteller. On Equality of Educational Opportunity. New York: Random House, 1972.

Myrdal, Gunnar. Objectivity in Social Research. New York: Pantheon Press, 1969.

Polier, Justine Wise. "Myths and Realities in the Search for Juvenile Justice," *Harvard Educational Review*, Vol. 44, No. 1, 1973.

Scarr-Salapatek, Sandra. "Race, Social Class and I.Q." *Science*, Vol. 174, No. 4016, 24 December 1971.

Schneirla, T.C. "Behavioral Development and Comparative Psychology," *Quarterly Review of Biology*, 41 (3), 1966.

Shockley, William. Letters to the Editors, *Scientific American*, Vol. 224, No. 1, January, 1971.

Singer, Mark. "Twelve Students Suspended in 'Shockley Affair,'" *Yale Alumni Magazine*, June, 1974.

Skodak M. and H.M. Skeels. "A Final Follow-up of 100 Adopted Children," *Journal of Genetic Psychology*, 75, 85-125, 1949.

Stoddard, George D. The Meaning of Intelligence. New York: MacMillan, 1943.

U.S. Industrial Relations Commission Final Report and Testimony Submitted to Congress by the Commission on Industrial Relations Created the Act of August 23, 1912. Two Volumes. Washington: Goverment Printing Office, 1916.

U.S. v. City of Jackson, Mississippi, 318 Fed. 2nd 1, 5-6 (1963).

PADFIELD-HERRNSTEIN AND LAYZER-JENSEN

James C. King

The present controversy is essentially one between those who espouse extreme genetic determinism in which there is a one-to-one relationship between genotype and phenotype—both morphological and behavioral—and those who see the phenotype as the result of a complex interaction between genes and environment giving every genotype a range of expression.

We see this dichotomy in the interpretations of the interactions between the educational system and the labor market, as described by Harland Padfield in "Social Marginalization" and as seen by Richard Herrnstein in a review of his book written by Hans J. Eysenck:

> It might be thought that what Herrnstein has to say is controversial or novel, but as he points out himself, it is nothing of the kind. Intelligence can now be measured with some degree of accuracy; intelligence is important in modern life because it is highly correlated with achievement, socio-economic status, and income; intelligence is largely inherited, although environment inevitably also plays a part; as our education becomes more egalitarian, so the importance of heredity in causing differences in I.Q. will increase. All of this was commonplace in academic psychology a quarter of a century ago, and when I made the same points in my first paperback, *Uses and Abuses of Psychology*, nobody blinked an eyelid. Herrnstein is perfectly correct in stating that there is no "controversy;" more recent evidence, which he quotes, has supported strongly the older views and has made more precise our models and our

qualitative estimates. In the circles of those who work professionally in this field, these truths are indeed taken as self-evident; it is politically motivated agitators, without any professional background in this field, who make the noise interpreted as "controversy." If there were any serious criticisms of the genetic argument, then one would expect to find them in the pages of the professional journals devoted to these issues; any such criticisms are conspicuous by their absence. The reader of Herrnstein's book will get an excellent survey of the field of intelligence testing, orthodox in every way, and without any attempt to dispute the existence of weaknesses where these exist, but also without the desire to gloss over socially undesirable consequences where these can be found. There are occasional points on which one might criticize the text, but these criticisms are unimportant to Herrnstein's main point.

This main point, as I take it, is that we should give serious thought to the social consequences of the known facts about intelligence. A simpleminded egalitarianism has overemphasized the importance of environment in creating a more just and equal society; we must look at the limitations which biology has placed in the way and consider what can be done to make society a better and more just place in which to live. Herrnstein points out certain consequences in the inheritability of the I.Q.; these may be desirable, but that does not mean that they will not come about. Nor does it mean that we should blame Herrnstein for pointing them out to us; the custom of ancient tyrants who slew messengers who brought ill tidings does not sit well on modern academics! I suggest that we pass a vote of thanks to Herrnstein and read his book; hopefully this will start a debate better informed, and to better purpose, than that which he recounts in his first chapter.(1)

Padfield sees the operation of a complex system through which the young are processed toward employment; as they move through the system, the young are assigned differential credentials which assign to them status consistent with market demands. To Herrnstein, the same process simply illustrates how individuals adjust themselves in obedience to their genetic endowments.

In "Behavioral Science and Society," Layzer has emphasized the long and tortuous road from gene through product to morphology and similarly from morphology to behavior. It is these complexities which account for the range of expression or norm of reaction as it is sometimes called. Ironically, it is the experimental geneticists like Lerner, Dobzhansky, and Lewontin who emphasize this flexibility; genetic determinism is more popular among the psychometricians.

Time after time, one finds the psychometrician automatically accepting a genetic explanation without considering an alternative. When Jensen, as he relates in the preface to *Genetics and Education*, found that culturally deprived children with low I.Q.'s were more competent in playground society than middle class children of corresponding I.Q.'s and could, in fact, perform as effectively on specially constructed tests as average or even gifted middle class children, he immediately invoked "independent polygenic factors" to explain this difference in behavior. He does not seem to have considered the possibility that the educationally mentally retarded children of the culturally deprived group might do poorly on standard tests because of their conditioning.

An illustration of a similar predilection for a genetic explanation, even in spite of contrary evidence, is provided by a quotation from the late Sir Cyril Burt, whose study of twins contributed as much as any single work to the belief in the figure 0.8 for the heritability of the I.Q. In an address given in May, 1957 at the University College in London(2), Sir Cyril stated:

> The figures show that the abler children from the working classes, even when they have obtained free places or scholarships at secondary schools of the "grammar" type, frequently fail to stay the course; by the time they are sixteen the attractions of high wages and of cheap entertainment during leisure hours prove stronger than their desire for further knowledge and skill, and easily overcome their original resolve to face a long prospect of sedentary work in *statu pupillari* . . .Underlying all these differences in outlook, I myself am tempted to suspect an innate and transmissible difference in temperamental stability and in character, or in the neurophysiological basis on which such temperamental and moral differences tend to be built up. Tradition may explain much; it cannot account for all.

Sir Cyril's statement does not reflect a statistical error; it is not even a statement of science.

NOTES

(1) The New York *Times*, September 1, 1974, p. 15. This excerpt was originally published in Books for Libertarians, December, 1973. In the *Times*, it, as part of a review, appeared in an advertisement, "reprinted as a public service with permission of the reviewer," by the Foundation for Human Understanding, 1225 Connecticut Avenue, N.W., Washington, D.C. 20036.

(2) *The American Psychologist*, Vol. XIII, 1958.

THE SCIENCE INFORMATION MOVEMENT: NON PARTISAN BUT NOT NEUTRAL

Evelyn A. Mauss

Abraham Edel identified as one of the problems of the scientist's responsibility "the mode of publicizing results and the ways of engaging in controversy about them." This problem spawned the science information movement in the United States, of which the New York Scientists' Committee for Public Information (SCPI) was an early manifestation. SCPI communicates scientific information, i.e., "publicizes results," to an array of people and groups, including trade unions, school children, church groups, parents' associations, television and radio audiences, and local, state, and federal legislative bodies, but the scientists who speak for SCPI do not take policy positions. SCPI hopes to develop an electorate which can itself make informed policy decisions on issues with science content (and, when the electorate are also legislators, to apprise them of the facts relevant to specific legislative issues). The effectiveness of the work of SCPI has yet to be proven, since measurement and evaluation in this area are primitive.

Current SCPI task forces include one related to lead poisoning in children (vide Allan Chase's paper on pellagra, and the confident description of its victims as "slow, lazy, etc."), and another related to the biology and sociology of race. The history and performance of SCPI's task force on race provide a case history of public information activity.

As we know, the current controversy about race and intelligence has roots that reach back to colonial days. Since the matter of race and intelligence was perceived even then as a science question and also as a question, to which answers could influence public policy on slavery, it is not surprising that the matter provides a prototype for the contemporary public information movement.

The Abolition Society of London had requested the Philadelphia Society for the Abolition of Slavery to collect and document instances that exemplified the intelligent behavior of Negroes. The Society sought to amass and utilize evidence on the equality of the black races to the white, with respect to their mental abilities, and thus to support the antislavery movement.(1)

Among the American scientists and physicians responding to this request was the renowned Benjamin Rush. His communication concerned Dr. James Derham, a Negro physician of New Orleans.(2) Dr. Rush stated facts and drew no inferences from them. He told a little of the early life and medical training of James Derham, but described how, at the close of the Revolutionary War, Derham was sold to Dr. Robert Dove of New Orleans. Dr. Dove "employed him as an assistant in his business, in which capacity he gained so much of his confidence and friendship that he consented to liberate him, after two or three years, upon easy terms." Dr. Rush then described some of Dr. Derham's education and activities, and continued, "I have conversed with him upon most of the acute and epidemic diseases of the [section] where he lives, and was pleased to find him perfectly acquainted with the modern, simple mode of practice in those diseases. I expected to have suggested some new medicines to him, but he suggested many more to me.... He speaks French fluently and has some knowledge of the Spanish language." Dr. Rush was happy to call Dr. Derham "a brother in science" and to commend him to the Pennsylvania Abolition Society as evidence of Negro achievement.(3)

How do we assess the utility of that early American public information effort? Thomas Jefferson, who was Rush's friend and colleague, said that he "read with delight every thing which comes from your pen." Certainly, Dr. Rush must have reinforced Jefferson's hesitant inclination to regard Negroes as inherently equal in intelligence. Certainly, Jefferson was tortured and ambivalent and never resolved the problem for himself, although he tried to be objective:

> The opinion, that they are inferior in the faculties of reason and imagination, must be hazarded with great diffidence. To justify a general conclusion, requires many observations, even where the subject may be submitted to the Anatomical knife, to Optical glasses, to analysis by fire, or by solvents. How much more then where it is a faculty, not a substance we are examining; where it eludes the research of all the senses; where the conditions of its existence are various and variously combined; where the effects of those which are present or absent bid defiance to calculation.

Nevertheless, having met several of Dr. Edel's criteria, and having recognized the "heredity vs. environment" pitfall, Jefferson's conclusion still was: "It is not their condition [i.e., their environment] then, but nature, which has produced the distinction."(4) A dilemma persists: when information is publicized by scientists, how can we assess its effect in a sociopolitical context?

To return to the contemporary heredity-environment behavior controversy and the responsibility of the scientist: The geneticist, the late L.C. Dunn, who referred to the error of some early views of race uniformity, race purity, and fixity of racial differences, wrote that

> ... we should as reasonable beings like to believe that if we get rid of our biological misconceptions, we should thereby cure the social and political ills and injustice and exploitation which appear to be based upon wrong biology.(5)

There was scientific information on race which the public ought to know, and SCPI added the subject to its agenda in the sixties, before Jensen published his *Harvard Educational Review* paper in 1969.

Progress was slow and audiences were often hostile. Only junior high school and high school students were consistently interested and responsive. They asked, "Why don't we get this stuff in school, instead of in a biology club lecture in the evening (or occasionally in honor society or Problems-in-American-Democracy meetings)?" "This stuff" was straightforward biology: the meaning of race, the development of human differences, the meaning of a controlled experiment, and the question of race and intelligence. Harry Milgrom, the Director of the Bureau of Science of the New York City Board of Education, was in a key position to determine the content of the science lessons that his one million, one hundred thousand charges were taught, and he was interested in a small study unit on the Biology of Race. SCPI prepared a course of study, and it was field-tested, revised, and eventually published as the Board of Education's first minicourse.(6) During the early 1970's, SCPI also worked with the Board of Education's television channel eventually to produce three videotapes based on the study unit.(7) Each videotape was scheduled for three or four broadcasts a week, and the minicourse was distributed to science chairpersons in junior and senior high schools throughout the city. The information in both was scientifically rigorous, and we hoped that the paucity of material on biology of race might be rectified. However, SCPI effort to teach this subject was sobering, because even though the videotapes and minicourse were valuable, the New York City Public School system was unmanageable and unaccountable. Consequently, informative programs were created and telecast hour after hour and day after day, but the classes for which the programs were created did not have, could not find, or never were issued receivers. Principals and teachers who feared controversial material or lacked commitment and conviction, tended to avoid a sensitive and difficult subject. From this experience and others in SCPI history, the conclusion is inescapable that third party transmittal of information may be self-defeating.

Evaluation of a program, even under optimal conditions, is still a serious problem. Whether audiences are parents or students, producers or consumers, legislators or trade unionists, we realize uneasily that the measurement and evaluation of effectiveness are nonexistent. For the biology of race project, we have received only anecdotal evidence of salutary outcomes; a carefully conceived attempt to conduct a proper program of evaluation was flouted by the very institution that encouraged its planning.

SCPI's effort to publicize scientific information on race dealt with materials organized and presented not necessarily by the scientists who did the research, but by scientists generally. The guidelines for publicizing this kind of secondary material are the same as for original work. According to Professor Edel, the material presented should

> ... somehow make clear all the kinds of issues involved [logical, conceptual, methodological, evidential, cultural, politico-social], together with the paucity of definitive evidence, the tentativeness of the conclusions and the present disagreement among experts.
>
> It should separate issues rather than bundle them in one package.
>
> It might well recount the history of the problem and its ideological entanglements, and suggest what scientific developments (e.g., in population genetics) have raised fresh hopes ... for new advance.
>
> But it should show the difficulties as well as the hopes in the investigation, and the importance of the presuppositions concerning the meaning of "intelligence" and "race" that underlie the formulations of the problem.
>
> Above all, perhaps, from the point of view of the history of science, it should consider the possibility that the formulations of the question will themselves undergo change as science advances, and raise questions whether present formulations may not be an inadequate basis for present investigation.

These guidelines have been followed reasonably well by SCPI presentations on the Biology of Race and on Race and Intelligence, but differently for different audiences.

Finally, Professor Edel states, "... there is no place for the dogmatism that jumps to a reconstruction of policy in the practical fields as if it were a deductive conclusion from an established thesis." One justification or "sound element" does exist in the inclination of some scientists—including those in the science information movement—not to take political or policy positions. That "sound element" is "recognition of the gap between science and social policy. It is a gap that is filled by other knowledge than the findings of the particular scientist, and by values and value-judgments in the community."

The general public must retain the prerogative of applying to scientific information its values and value-judgments. Abraham Edel suggests that allowing non-scientists to judge scientific information is an abrogation of "specific social responsibilities." On the contrary, it is the *assumption* of specific social responsibility for scientists and their organizations to undertake a methodical effort to provide information to the public, although they do not essay to fill "the gap between science and social policy." Scientists may have a concomitant responsibility to explicate the values and value-judgments that are involved in decision-making, but in attempting to distinguish science from public policy and science from ethics, the science information movement believes that the community (including scientists in it) itself must be the decision-maker.

Scientists who reject a non-partisan position can, as scientist-*citizens*, function anywhere along a continuum from ivory tower isolation, through a legislation-oriented middle (e.g., Federation of American Scientists), to high-profile activism. According to Gerard Piel, the removal of a case "out of the orderly discourse of science into the political arena" can have counterproductive repercussions. As we have witnessed, the demagogues, the elitists, the racists have in their vehement discourses on race and intelligence had a frightening influence on the controversy. Dissent cannot, should not be silenced but while scientists probe the questions as to the legitimacy, the effectiveness, and the appropriate routes of their invasion of the political arena, they can, as scientist-citizens, participate determinedly in the decision-making process. Public information activity purports to inform non-scientist-citizens who will make decisions and act. Scientists as citizens can do no less.

NOTES

(1) Butterfield, L.H., ed. *Letters of Benjamin Rush I*, 1951, p. 498 (editor's note).

(2) *Ibid.*, p. 497.

(3) Jordan, Winthrop D., *White over Black. American Attitudes toward the Negro, 1550-1812.* Univ. of N. Carolina Press, Chapel Hill, 1968, pp. 448-449.

(4) *Ibid.*, quoted on pp. 438-449.

(5) Dunn, L.C., in *Race and Science: Scientific Analysis from UNESCO.* Columbia University Press, N.Y., 1961, p. 13.

(6) Scientists' Committee for Public Information and Bureau of Science, Division of Educational Planning and Support, N.Y.C. Board of Education. *Biology of Race, A Five Lesson Mimicourse*, 1974.

(7) N.Y.C. Board of Education. *WNYE Channel 25 Television Manual*, 1973, pp. 5, 292-4.

BEHAVIORAL SCIENCE AND GENETIC DESTINY: IMPLICATIONS FOR EDUCATION, THERAPY, AND BEHAVIORAL RESEARCH

Ethel Tobach

> *He (Eichmann) hit me with his long hand. I still remember his hands, long and thin. And he said to me, 'Du schweine, du hund. You dirty dog, you dirty pig, you had enough brains to escape, so now you can live.'*
> From an oral history record being made for the American Jewish Committee, reported in the New York *Times*, May 11, 1976

There may be many who object to the suggestion that their concepts of the relationship between heredity and behavior, between natural selection and reproductive fitness, or between psychological theories and societal practices have any relevance to the quotation above. The setting for the story implied in the quotation is Nazi Germany and Eichmann administering Auschwitz, but it is universal for all societies in which some were slave-owners and some were slaves. In an editorial in *Nature* one is reminded of the need to explore the relationship between the theories of behavioral science and social policy: "It is difficult to escape the conclusion that those who go on saying that heritability is high are feeding racist tendencies, however inadvertently and seemingly innocently, and are supplying an apparent scientific basis for those who wish to justify racial discrimination."

This book is about the communication process that could have clarified the relationship between the activities of scientists and the consequences of those activities in society. Such a process is complicated, and it is most necessary for the behavioral scientist to understand it and be able to work with it. In the area encompassing heritability and behavior in the United States one of the most significant aspects of that process is to understand why some speak and act on the topic and some do not. Because of this tendency, the information we have about what behavioral scientists think about the issue is very specialized, and it is difficult, therefore, to understand how to communicate with all of those working in disciplines involved: educationists, anthropologists, psychiatrists, sociologists, psychologists, social workers, and many others who make decisions daily about the potential of individuals to change their behavior.

As Proshansky relates, the SPSSI Council was primarily motivated by that desire to establish a communication process among scientists who were in profound disagreement about the concept of the inheritance of behavior. Some of the people who voted for the establishment of the Commission were concerned that scientists with prestigious credentials were affecting public and academic opinion so that the likelihood of "tracking" members of minority ethnic groups into non-academic educational programs would increase. They were concerned that governmental educational policies would be affected and that an irreversible, deleterious effect would be to prevent minority groups and women from attaining the class mobility they desired.

Those who supported the hereditarian viewpoint apparently reacted to those concerns and were most outspoken about the use of the words "renewed assault on equality." Subsequently, in the interests of maintaining or initiating a true discussion, the name of the Commission was changed. One wonders if the change was warranted. Two matters preceding the appearance of "Comment," among others, should be noted.

The first is a press release dated October 16, 1956, which announced the issuance of a joint statement "attacking as 'scientifically unjustified' theories that the intellectual potential of Negroes is inferior to that of Whites... In connection with the process of school desegregation and the difficulties with which it has been accompanied in certain areas, the question has again arisen as to the existence of innate differences in intelligence between Negroes and Whites." The statement was signed by Otto Klineberg, Columbia University; Theodore Newcomb, University of Michigan; Gardner Murphy, Menninger Foundation; Nevitt Sanford, Vassar College; Robin Williams, Jr., Cornell University; David Krech, University of California; Jerome Bruner, Harvard University; Allison Davis, University of Chicago; Daniel Katz, University of Michigan; Anne Anastasi, Fordham University; Stuart Cook, Isidor Chein, Marie Jahoda, New York University; Kenneth Clark, College of the City of New York; Bingham Dai, Duke University School of Medicine; Irving Lorge, Teacher's College, Columbia University; Solomon Asch, Swarthmore College; and David Rapaport, Austen Riggs Foundation.

Twenty years after the historic decision of the Supreme Court was defended desegregation is once again an issue for science and society. The commission may have been misnamed in that the onslaught was not "renewed" but "continued." In the body of the press release, reference is made to the 1950 statement of UNESCO in Paris in which the following sentence appears: "In short, given similar degrees of cultural opportunity to realize their potentialities, the average achievement of the members of each ethnic group is about the same."

The United Nations some twenty years after that statement also felt called upon to deal with the issue. A United Nations Seminar on the Dangers of the Recrudescence of Intolerance was held in Nice, France, between August 24 and September 6, 1971. The full title of the United Nations document includes "the search for ways of preventing and combatting it." It would seem that the concern of the SPSSI Council was shared by others, and the Commission was not out of the mainstream of contemporary thought.

However, as in every other aspect of society, the mainstream is composed of many currents, and it is clear that today the dominant current is that of hereditarianism. It is therefore important to examine the statements made by the signers of the resolution which appeared in "Comment." Many of those have been dealt with by Proshansky. Two of the concepts in "Comment" and in a letter, dated January 9, 1973, by Lloyd G. Humphreys to Dr. Janet Spence, are also worthy of discussion.

The equation of "biology" with "hereditarianism" is clear in "Comment" in such phrases as "hereditarian view, or to recommend . . . a study of the biological bases of behavior;" "biological explanations or efforts;" "role of inheritance in human abilities and behaviors;" and "biobehavioral reasoning." Only one phrase might be inferred to have another viewpoint: "biological hereditary bases of behavior as a major complement to the environmental efforts at explanation." The phrase is not quite a statement of interactionism, but might be read as such. Those who do not subscribe to such views of behavior are "liberal" and "committed environmentalist."

Two issues are of importance here. The first is the use of the word "biology." Biology is usually defined as the study or science of life. "Life" is derived from the Teutonic root "to remain," from the Sanskrit "to anoint" and the root from "leip" from the Greek "to anoint." (Skeat, 1968) Advances in modern genetics have given a new definition of life which is based on the transmission of particular biochemicals leading to replication of forms. It is conceivable that this new conceptualization permits us to equate "biological" with "genetic" and "hereditary." In that sense, the meaning of "biological" bases of behavior should be understood to relate all aspects of behavior to genetic processes.

It is pertinent to understand the genetic process and how it relates to behavior. Much has been written on the genetic process (Ebert & Sussex, 1965; F.J. Gottlieb, 1966; Hirsch, 1970; Lewontin, 1970, 1974; Lehrman, 1970; Schneirla, 1971; Tobach, 1972, 1973) in its developmental aspect and in regard to behavior. There is a vast literature which is difficult for the non-geneticist to master and keep up to date. It is clear that certain principles have been established that are critical for the behavioral scientist, however. First, the gene can only be defined as it expresses itself in a given milieu. At all times, the characteristics of the gene are understandable only when the conditions in which it functions are defined. A corollary of that principle is that when one studies particular traits, it is not always possible to know which definable milieu variables are relevant to the expression of the gene. Determining relevancy requires careful and repetitive experimentation frequently. Thus, much genetic research concerns itself with elucidating the conditions which make it possible for genes to function, to be inhibited, or to have their activity delayed. Botanical investigations give striking examples of organisms carrying the "same" genes which vary in color or other characteristics when they have been grown in different temperatures or levels of humidity. (Wallace & Srb, 1961; Wallace, 1968)

When one studies the expression of genes by looking at biochemical and structural variations, the direct relationship between genes, a biochemical level of organization, and other biochemical and biophysical aspects such as pigment in a petal, or electrophysiology in a neurone, or response of neurones to sugars is clear. (Ikeda & Kaplan, 1970) The indirect pathway from biochemical levels of functioning to behavioral levels of functioning is difficult to chart. (Teyler, Baum & Patterson, 1975)

An example is the study by David Bentley (1975) in which he shows the chemical effect of ethyl methanesulfonate on neural structural development. Treating crickets with this chemical resulted in changes in self-grooming by animals and withdrawing from a puff of air. Both these behavioral traits could be manipulated by various breeding procedures to "demonstrate" that the behavior is inherited. Actually, the chemical affected the development of synapses between nerves and muscles, and the development of hairs on the body. Animals lacking these structures and their functional relationships showed changes in complicated behavior patterns called "grooming" and "evasion."

As Kuo (1967) pointed out, the understanding of the relationship between events on the biochemical level of the gene and the development of neurones and other structures that do not specialize in a species-typical fashion is only part of the story. The critical process is to intervene at different developmental stages and change the course of events in order to understand how the absence of some structure can be overcome by the development of other structures or functions which might serve the organism. It is a different view of genetics, biology, and behavior than that which those who talk of heredity and its limitations offer us.

Second, the opposing of environment and heredity in "Comment" is not necessary. The concept of levels of organization and integration, fused with the developmental approach resolves the apparent dichotomy between genes and experience, heredity and environment, instinct and learning, or any other of the pairs of contradictories posed for behavioral scientists. Many have seen this contradictory apposition as a non-problem (Hinde, 1970; Schneirla, 1971; Lehrman, 1970). A developmental approach deals with all levels of experience, organization, and integration on physical, biochemical, genetic, physiological, social, and other levels.

Another use of the term "biobehavioral," or "biological," confuses physiological with genetic processes. Biological is seen to equal physiological. This is an interesting development in the use of words; physiological as discussed by Skeat (1968) says that Blount's *Glossary* used the term to mean an inquiry into the nature of things; again, if one accepts the simplistic version of genetic processes, one may see that biological, which is equal to physiological, results in physiological being equal to genetic. This indeed, is an interesting example of syllepsis.

These concepts of the relationship between biology and genetics are the most widely held in the life sciences, and reflect a general acceptance of the hereditarian view of all physiological and behavioral phenomena. This relationship has led, despite the statements in "Comment," to a continuance of research in the relationship between genes and behavior, and indeed an increase in research and publications. Articles dealing with "behavioral genetics" of animals other than humans appear not only in *Behavior Genetics*, but in *Science, Nature, Animal Behavior, The American Naturalist, Proceedings of the National Academy of Science, Bulletin of Psychonomic Science*, and many other journals. A listing of the articles specifically related to intelligence, heredity, and race would require a book in itself. However, the partial listing incorporated into the bibliography at the end of this paper indicates that research in other aspects of human behavior and genetics is flourishing. Further, this listing reveals that the concept of "race" is used generally and that comparisons of many different ethnic groups in regard to physiology and other characteristics have not been suppressed.

Is the biobehavioral viewpoint expressed only in "basic" research, or does it get translated into educational practices as well? The economic expression of the hereditarian view will be cited later, particularly with reference to advanced degrees. One case of its expression in medical education is currently receiving attention. In the September 26, 1975, issue of *Science*, Bernard D. Davis takes issue in an editorial with a NOVA program (a series about the frontiers of science which is broadcast on public television primarily) in which "a distinguished population geneticist denied the legitimacy of human behavioral genetics." There were responses to that editorial in subsequent letters to the editor and in the pages of *Science* in the form of scientific articles. At a subsequent meeting of the American Association for the Advancement of Science in February, 1976, in Boston, Dr. Davis repeated his comments with documentation in a traditional scientific presentation. At that time, he raised the issue of the lowering of medical school admission standards because of the pressure of minorities and women to admit more people from those groups. Subsequently, in an article in the New York *Times* on May 13, 1976, Dr. Davis is quoted on the basis of an article he wrote for *The New England Journal of Medicine* in which he criticized "substandard academic qualifications." He added that it was "virtually impossible to flunk out even the worst white students" because "exceptions were made in passing minority group students." Dr. Davis is a leading medical educator at Harvard.(1)

The Harvard Medical School administrators immediately denied in the *Times* of May 19 that minority students obtaining degrees from their school were any less qualified than other students. A spokesman for the Association of American Medical Colleges also refuted Dr. Davis' statement in this article which went on to state that a good deal of discussion had occurred among the medical professionals about the lowering of standards. The move for a relicensure procedure of physicians after they have left medical school is being supported and planned by medical professionals. The controversial statements by Dr. Davis may yet result in a truly beneficial examination of the competence of the entire profession. Reexamining physicians who have been in practice many years, regardless of their ethnic background or sex, would comfort the receivers of medical service.

Another serious aspect of the acceptance of the concept of hereditarianism by educators and their teaching of that viewpoint with apparently no awareness of its controversial and questionable scientific validity is seen in the curricula developed with the support of the National Science Foundation for pre-college science students. Three examples of this phenomenon are of interest.

The program undertaken by the National Science Foundation was designed to do several things. It aimed to upgrade the science background of students in kindergarten through the twelfth grade who were interested in becoming scientists. It attempted to encourage minorities and women to enter science. Further, it aimed to educate non-scientifically oriented students in order to create a public who understood the scientific method and would be able, therefore, to understand the work of scientists and its value.

In the course of this program, a curriculum called "Man: A Course of Study" (MACOS) was developed. In an article in *Social Education*, prepared for the Science Curriculum Implementation Review Group of the Committee on Science and Technology, U.S. House of Representatives, Peter B. Dow discussed the material and rationale of MACOS. The organizers of the project, one of whom is Jerome Bruner, follow the traditional approach to the comparative study of the evolution of behavior. The curriculum covers the life cycle of salmon, the behavior of herring gulls, baboons, and finally, the Nesilik Eskimos, as a means of studying the special characteristics of human behavior. The section about herring gulls is entitled "Innate and Learned Behavior in the Life Cycle," and is based on the work and theory of Tinbergen and Lorenz.

In this book of cartoons and captions entitled "Innate and Learned Behavior," readers encounter the following:

In Panel 13, the caption for a drawing of a herring gull's brain, states, "Yet the chick has had no experience with this set of behaviors in the past. How does it happen to carry out these behaviors? The herring gull chick is born with instructions for carrying out certain behaviors. These instructions are already built into the brain." The panel shows a brain of a chick with compartments labelled "pecking" and "eating"—a conceptualization about brain compartments which is controversial. (Valenstein, 1975) In Panels 19 and 20, the captions state, "The kind of automatic behavior that is ready at birth is called *innate* behavior. Past experience is not necessary for innate behavior." Here, the view is an orthodox one which Hinde (1970) refutes in his book. In Panels 33, 34, 35, and 36, the captions read, "All animals are born with some innate behaviors. Fingerling salmon swim upstream. Baboons cling to their mothers. And human babies cry when they are hungry." This presentation of innate behavior disallows any other concepts about the relationship of these species to one another. In Panels 60, 61, 62, and 63, the captions for a drawing of a developing brain of a bird states, "As the animal's body grows, instructions for new behavior develop in its brain. Then the innate behavior appears at the proper time in the animal's life." A section of the bird's brain, which is shown growing larger as the bird grows, is labelled "egg sitting." In Panel 63, we see the bird standing over an egg. The presentation of the relationship between the nervous system and behavior is erroneous and again ignores the complexity of the development of reproductive function in birds. (Lehrman, 1961)

This material is being used in some 1400 schools in the United States; at least in one instance it was being used in a very expensive private school in the kindergarten class, although it was intended for a higher grade. The theme of the book is developed around the concept that in the lower forms, innate behavior is in the ascendancy, and in human beings learning is predominant. The objections to this curriculum came primarily from a conservative group of parents who were religious fundamentalists.

Scientists and others defended the MACOS project on the basis that most criticism came from people who generally did not support scientific endeavors, or federal planning, and in some instances were even opposed to the teaching of evolution in particular. A review of the many articles which appeared in the press, or the proceedings before the Congressional Committee on Science and Technology, reveals that there was no objection to the teaching of ethology to the students in the MACOS program.(2)

Two other curricula which were not yet in use were similar to MACOS: *Exploring Human Nature* and *Understanding Human Behavior*. Both of these were developed by famous scientists in their fields. The *Understanding Human Behavior* curriculum was the product of a prestigious committee established by the American Psychological Association. This curriculum as it was seen by the panel constituted by the National Science Foundation in December, 1973 was based primarily on the dichotomous concept that behavior is either instinctive or learned. Sometimes this pair of adjectives is termed "psychological" and "biological."

Exploring Human Nature was the work of Irven DeVore, George W. Goethals, and Robert L. Trivers. Trivers (1971) is a theoretician in sociobiology, the keystone of which is the concept of altruism. The basic process in evolution is viewed as gene transmission; individuals usually behave in a fashion to guarantee the greatest degree of transmission of genes from the gene pool to which the individual belongs. Sometimes such behavior leads to the destruction of the individual; this behavior is called altruism. (Wilson, 1975)

The units in the program speak for themselves:

> Unit 1: Origins of human behavior; how natural selection has shaped the behavior of our species; similarities in behavior across cultures probably point to human biological universals; different forms of the

human family can be explained in terms of past natural selection pressures; human beings are the way they are today because of past interaction of individuals with their environment.

Unit 2: Childhood and the community; the biological bases of human behavior limit what is possible in human communities; how different social behaviors made sense when viewed in the context of the whole community; the form the family takes influences and is influenced by other aspects of the community; the structure of the community influences the behavior and experience of the individual.

Unit 3: Coming of age; managing transitions; biological changes at puberty have important social and psychological effects upon young people; the pattern of transitions from child to adult reflects the structure of the society; social structures influence whether or not the individual is responsible for his or her own transition to adulthood.

Unit 4: The individual in society; competition and cooperation; in every society, factors of age, sex and kinship shape adult roles; the choice of adult roles and the opportunities for competition, cooperation and achievement are different in different societies; in every society the family is the basic unit of economic cooperation and its members perform specialized tasks; how members of a society conform to or change its patterns of competition and cooperation.

The bias in the theoretical constructs in the genetically determined gender role is clear throughout. In Unit 1, the *Origins of Human Behavior*, the children are asked to simulate a natural selection experiment, using human characteristics. It is not surprising that the panel reviewing the curriculum criticized this unit as an exercise in eugenics.

Exploring Human Nature is a very handsome product of audio-visual programs, teachers' guides, and books for children. Throughout the materials there are explicit statements about human nature based on Lorenz, Tinbergen, Eibl-Eibesfeldt and E.O. Wilson. The view presented is that behavior is inherited, and experience can only modify the pre-programmed potential in the genes within given restrictions. Learning processes for humans and other animals differ, it is explained, because human beings are prewired to learn as no other animal is able to learn, and human behavior differs from animal behavior in the learning aspect but not in the instinctive aspects, including aggression and reproduction.

The panel itself did not object to the inclusion of material based on the concept of inheritance of behavior; the usual criticisms of the entire National Science Foundation project were heard: dissatisfaction with the cross-cultural approach; the possible invasion of the student's privacy by behavioral study techniques; the manipulative aspect of the teacher-student relationship. While these criticisms have some merit and should be explored by behavioral scientists and educationists, the fact is that genetic destiny is being taught to children from the time they enter the school system.

The teaching of a hereditarian viewpoint at all educational levels takes place elsewhere as well. In the Spring, 1973 issue of *Biology and Human Affairs*, the editor comments on the spreading study of social biology in Great Britain. "The appearance of an examination syllabus in Social Biology at 'A' level is an event of some significance, which takes its recognition as an acceptable subject a stage further than that already signified by its establishment and spread in university and college of education circles."

The theoretical emphasis of this syllabus is described: "The third subsection on Heredity and Human Diversity can clearly be integrated with the second for there is an evident mandate for a thorough study of genetic matters." (p. 2)

Later (p. 4), "the human use of social biology" is described. "Social biology scrutinizes too social studies...sometimes the models of economic or sociological institutions are seen to be oversimplifications by the application of observed human innate variation to them—a danger which the syllabus has recognized."

It is hard to reconcile such developments among educationists with the statements of the hereditarians in "Comment."

The dynamics responsible for the renewed reliance on hereditarianism as a valid scientific concept are usually clearly seen in the state of economic health and societal tension, as Edel, Piel, and Padfield demonstrate. There is another aspect of the economic state of society that directly affects scientists: the willingness of society to support work that would on the surface appear to have little or no immediate value to society. A low priority in this respect (Tobach, 1972) is more likely to be assigned to behavioral science than to chemistry and physics which are important to engineering and technology. These have first priority, and medical science is second. Behavioral science only pays its way when it does things that have *prima facie* significance: testing chemicals that affect behavior that may be used in military operations; devising techniques to be used in propaganda or other instruments for changing human thinking; and providing rationales for values that are important to the *status quo* of the society.

Some recent events in the American Psychological Association Council of Representatives meetings are an example of the complicated processes involved in this area of discussion. At the meeting in August, 1975, a resolution was introduced, which, if passed, would have required the Association to ask that all credentialed psychological tests bear a statement cautioning the testee and the tester against possible misuse and abuse of the test. This resolution was tabled until the next meeting at which time it was sent to committee. At that same second meeting, in January, 1976, the Council passed a resolution which affirmed the usefulness of tests and upheld the Association's monitoring procedures.

One may interpret these events in many ways. It should be noted that attempts to combine the two resolutions were only partially successful. The incident is an example of the process that requires intensive and extensive examination of the possible investments that people may have in the hereditarian viewpoint. Geneticists whose life work assumes certain relationships between behavioral and genetic processes cannot be expected to easily give up their theoretical base. It is to their credit that Lewontin (1974) and Hirst (1975) among others show by their writings and work that their concepts of genetics are valuable and supportable by all sectors of the population.

Those who make a living from tests, personnel selection, and the other superstructural industries relating to society need to examine the relationship between their theories and practice. (Schwebel, 1975) Binet's original intention was to be of humanitarian use to the society that wished to educate its youth and those languishing in hospices for the socially disabled. Nonetheless, the outcome of those tests was to perpetuate an inequality of opportunity for just those populations for whom the tests were designed to help. In the years of testing millions of students and others, tests of the achievement of the student in order to diagnose areas of weakness were never developed for use with large groups. Furthermore tests to diagnose what was wrong in the educational process have never been developed for a mass basis. Test developers could construct achievement tests which would inform educationists and the public about what was lacking in the educational process. The recent concern about the poor performances on scholastic achievement tests throughout the country has taken the ugly shape of attempting to blame the failures on the integration of different ethnic groups into one school, on the recruitment of teachers, on programs of bilingualism, on innovative educational programs, and on other factors that reflect various discriminatory attitudes.

It is alarming that the testing industry has not been able to use its skills to help develop mass diagnostic tools. As Wallach (1976) and others have said, it is imperative that the test constructionists begin to examine the value of and the assumptions underlying tests. The need to assess the progress of mastery of skills and information and of solving problems is meaningful when the results are used to plan further training for the individual.

Tests are used primarily to determine the type of education to be given those who are tested. This use clearly affects the possibility of entry into the professions and highly skilled work that will be open to individuals who must demonstrate on tests their fitness for that type of work. Is it possible to relate the postulate that there are populations of human beings who are unable to deal with certain types of thinking by virtue of their genetic constitutions with a news story by Lesley Oelsmer in the New York *Times* of January 18, 1976? The Supreme Court is being asked to decide whether white males are being discriminated against by affirmative action programs. Shortly before that, in the September 5, 1975 issue of *Science*, Betty M. Vetter, Executive Director of the Scientific Manpower Commission writes that "A new report... shows that among almost 207,500 science and engineering Ph.D.'s in the U.S. labor force, 93.4 per cent are white and 92.1 per cent are male." In the New York *Times* of December 28, 1975, Gene I. Maeroff writes that the United States Office of Civil Rights has approved affirmative action plans for only 31 of the nation's 1,300 eligible institutions. There has been no monitoring of approved plans and no college or university has lost its federal funding, as the law provides, for failing to comply. Figures from the American Council on Education indicate that the presence of black faculty members on campuses has risen from 2.2% to 2.9% in 1972-73.

It is clear that the significant changes that many hoped would occur have not occurred. The hereditarians tell us that genes are responsible for the failure.

What happens to those who disagree with that analysis and wish to deal with the issue in scientific journals and other forms of communication? It is comparatively difficult to obtain information about these experiences, but three examples are possibly representative.

Chemical and Engineering News, in its January 13, 1975 issue printed an item which purported to support the conclusion that "... I.Q. depends more on heredity than on environment." (p. 48) Subsequently, a letter was sent to the journal by a group of faculty members from a prestigious polytechnical institution pointing out that "It is inappropriate and unscientific for *Chemical and Engineering News* to provide a forum for arguments presented without adequate documentation. Careless reporting of this potentially inflammatory subject must be deplored, particularly when alternative views are not presented." Their letter was not published, but instead the editor replied in a letter that "... the *Newscripts* page is not specifically devoted to purely scientific presentations. Its purpose is, on the contrary, to present enjoyable and amusing reading which has a direct interest to a scientific audience."(3)

An article by Berkeley Rice entitled, "The High Cost of Thinking the Unthinkable" appeared in the December, 1973 issue of *Psychology Today*. One of the people who was severely criticized in the article attempted to obtain equal space for a response, and to date the editors have not replied to the request.(3)

The behavior of the American Psychological Association in regard to the appearance of "Comment" also indicates the complexity of the social process by which scientists of different opinions communicate with each other. The Association, in response to the many demands by minority groups and women who considered themselves to be disenfranchised by the psychological profession and the scientific community, created first an Ad Hoc Committee, which then became the Board of Social and Ethical Responsibility in Psychology. At the February, 1972 meeting of the Ad Hoc Committee, the then Executive Director of the Association reported that an advertisement submitted by Dr. E. Page was being printed in the "Comment" section of the *American Psychologist* because of its importance to the Association and its members. At that time, a resolution was introduced to the Ad Hoc Committee that it ascertain whether psychologists had indeed been prevented from carrying out their scientific and professional pursuits. This resolution was defeated.(3)

Subsequently, when the SPSSI Council established the "Commission on the Renewed Assault on Equality" a group of the signers of "Comment" wrote a letter of protest to the Association's Board of Scientific Affairs, which then referred the matter to the Board of Social and Ethical Responsibility in Psychology. No further action was taken.

It is interesting to note that in the letter criticizing SPSSI's action, the use of the word "equality" in the title of the Commission is questioned. The writer of the letter states that the two meanings of the word "equality" have been confounded by SPSSI. He states that "the resolution in 'Comment' was one of persons entirely committed to political and social equality, that they wanted to introduce the scientific facts of biological and psychological individual differences intelligently and humanely into relation thereto." (sic) This objection was ill-founded, because the Commission was constituted to deal with equality only in a political and social context, both for the people who were being told they were intellectually inferior and for the scientists who were being mistreated.

One might question the source of the impetus to draft the resolution within "Comment." Certainly one of the most-significant sets of events was Dwight J. Ingle's letter to *Science* in 1965, his opening the pages of *Perspectives in Biology and Medicine* to the discussion of the race and intelligence issue, his autobiography in 1963, and his participation in a symposium on the concept of race at the American Association for the Advancement of Science meetings in 1967. (Mead, Dobzhansky, Tobach, Light, 1968) Another event that may have precipitated the writing of "Comment" was an exchange between Jerry Hirsch, an outstanding behavior geneticist and a journal regarding a series of criticisms and rebuttals of a statement that Hirsch made about Jensen. Hirsch asked for a publication of the exchange in the light of the important issues raised for behavior and genetic studies. When the journal refused, Hirsch made the material public within his university. The response to this publication was a telegram signed by many people who had signed the resolution.(3)

The viewpoints of hereditarians also receive protection indirectly. A French psychologist who had written an article critical of Jensen for *Psychologie* was told that his review could not be printed because it was too critical of a person as famous as Jensen.(3)

What can the behavioral scientist do to maintain healthy debates over substantive issues within the profession, and at the same time discharge a meaningful responsibility to society? Behavioral scientists have done less than other scientists. The most memorable and earliest example of scientific debate and social responsibility was the activity of the physical scientists in regard to nuclear energy and power. Ecologists have become so committed to their responsibility to society that they are now calling for an applied science of ecology. (Bingham, 1975) Geneticists have been seriously engaged in decisions about policing research efforts in areas that are deemed potentially dangerous to humanity.

Why has it taken the psychologist, among other behavioral scientists, so long to develop some means of engaging each other in significant discussions on race, intelligence, and educational practices? Friedrichs (1973) did a small survey in 1969-70 in which he asked a sample of people listed in the American Psychological Association directory whether they agreed or disagreed with Jensen's conclusions in the *Harvard Educational Review*. He found that most of the respondents disagreed or tended to disagree with Jensen. However, two factors affected the frequency of disagreement: the age of the respondents and whether or not the respondent lived in Alabama or Mississippi. The respondents' sex, type of specialization (physiological or clinical psychology) and being Jewish did not distinguish those who agreed or disagreed with Jensen. When we compare those respondents who belonged to such divisions of the APA as Tests and Measurements, Experimental, Educational, and School Psychology, with members of other divisions of APA, one finds an increase in the number of members who disagree with Jensen. It may be that the great majority of the psychological community does not agree with Jensen, but apparently the majority does not voice that objection.

The concept of some self-disciplinary machinery in the scientific community is not new. Many professional societies have traditionally monitored the behavior of their members, both for the good of the public and for the good of the profession. One organization that has been actively engaged in exploring the responsibility and accountability of scientists and professionals has been the American Orthopsychiatric Association (Shore & Mannino, 1975). The American Association for the Advancement of Science has also recognized the problem of accountability and has created a Committee on Scientific Freedom and Responsibility. (Edsall, 1975) For the most part, the most stimulating activities were carried out by dissident students and younger scientists, who effected change, particularly in raising the consciousness of the professions about their responsibilities to society. From time to time suggestions for specific institutions are published: Sieghart, in *Nature* in 1972 called for "an organized body of scientists" which should be "wholly independent of any kind of political or financial influence." This body would inform the public of scientific issues and events that are of significant interest to them . . . to "articulate the social conscience of . . . scientists." In constrast, a proposal by Kantrowitz (1975) which is still being discussed calls for a scientific advisory group of "court" appointed by the President (Wilford, 1976).

The problem of articulating and then implementing the social consciousness of the scientist through democratic means is confounded by the complexities of the instrument being used to solve the problem: the scientist. As Franz Boas said,

> Can we say conscientiously that scientists are not influenced by demagogues, catch-words and slogans? Is it not rather true that a great many of us, who may be clear thinkers in our own fields, are so little versed in public affairs, so much confined in our narrow field, that we are swayed by passionate appeals to outworn ideals or to a selfish interest that runs counter to the interest of the people? So when we speak of the need for education, do not let us forget that we have to educate ourselves.(4)

That process of education requires us to analyze those societal processes which govern the creative search for understanding of all aspects of life, including the genetic code and its function. We are also then required to understand how this quest relates to the social meaning of the scientist, and to the organization of social institutions. Finally, one must understand the relationship between these complicated expressions of humanity and the brutal, dehumanizing explication of the version of natural selection by which Eichmann made his decisions at Auschwitz. The extreme forms of oppression vary with different societies, and it behoves the scientist to ask how oppressive forces in society might use the results of scientific research to play the game of genetic selection with blacks and other minorities in our society.

NOTES

(1) On May 22, 1976, the New York *Times* carried Dr. Davis' apology for the article that he had written. He made it clear that he did not intend to cast aspersions on the qualifications of any of the medical students regardless of their ethnic background. He indicated that he had not realized that his article would be published in the public press. In a sense, this series of events is an excellent example of the issues facing scientists and academicians today: there is no ivory tower and the scientist faces the same forces as all other citizens in the marketplace. It is up to the scientific community to show how it can apply the scientific method to solving these problems.

(2) Although it is not clear from the listing in "Comment" what his responsibilities were and when they were discharged, Lloyd G. Humphreys is an active leader of the Academic Coalition in the American Psychological Association Council of Representatives and is listed as an Assistant Director for Education of the National Science Foundation. (1975, p. 95)

(3) All the materials and correspondence relating to this paper are on file in the SPSSI archives.

(4) Boas, Franz. Race and Democratic Society. J.A. Augustin Publisher, New York, 1945, p. 217.

REFERENCES

Altman, Lawrence K., Professor Contends Medical Schools' Standards Have Dropped Because of Rise in Minority Students. The New York *Times*, May 13, 1976, p. 14.

Bentley, David, Single Gene Mutation: Effects on Behavior, Sensilla, Sensory Neurons and Identified Interneurons. *Science*, 1975, 187, 760-764.

Bingham, Charlotte, Commentary—The Ecologists' Niche. *Ecology*, 1975, 56, 1-2.

Biology and Human Affairs, Editorial, 1973, Vol. 38, No. 2.

Davis, Bernard D., Social Determinism and Behavioral Genetics. *Science*, 1975, 189, No. 4208.

DeStefano, G.G., A Study of Morphological and Genetic Distance among Four Indian Villages of Nicaragua. *Journal of Human Evolution*, 1973, 2, 231-240.

Dow, Peter B., MACOS Revisited: A Commentary on the Most Frequently Asked Questions about Man: A Course of Study. *Social Education*, 1975, 39, 388-396.

Ebert, James D. and Ian M. Sussex, Interacting Systems in Development. Holt, Rinehart & Winston, Inc., 2nd ed. 1965.

Edsall, John T., Scientific Freedom and Responsibility. *Science*, 1975, 188, 687-693.

Exploring Human Nature, Brochure published by the Education Development Center, Social Studies Program, Cambridge, Mass.

Falek, Arthur, Differential Fertility and Intelligence: Resolution of Two Paradoxes and Formation of a Third. *Journal of Human Evolution*, 1972, 1, 11-15.

Friedrichs, Robert W., The Impact of Social Factors upon Scientific Judgment: The "Jensen Thesis" as Appraised by Members of the American Psychological Association. *Journal of Negro Education*, 1973, 42, 429-438.

Fuller, Renee N. and Joyce B. Schuman, Genetic Divergence in Relatives of PKU's: Low I.Q. Correlation among Normal Siblings. *Developmental Psychobiology*, 1974, 7, 323-330.

Goodman, Madeline J., Chin S. Chung and Fred Gilbert, Jr., Racial Variation in Diabetes Mellitus in Japanese and Caucasians Living in Hawaii. *Journal of Medical Genetics*, 11, 328-334.

Gottlieb, Frederick J., Developmental Genetics. Reinhold Publishing Corp., New York, 1966.

Hinde, Robert A., Animal Behavior. McGraw-Hill, New York, 1970.

Hirsch, Jerry, Jensenism: The Bankruptcy of "Science" without Scholarship. *Educational Theory*, 1975, 25, 3-28.

Hirsch, Jerry, Behavior-Genetic Analysis and Its Biosocial Consequences. *Seminars in Psychiatry*, 1970, 2, 89-105.

Ikeda, and Kaplan, Proceedings of the National Academy of Sciences, 1970, 66, 765-772 and 67, 1480-1487.

Itoh, S., K. Doi and A. Kuroshima, Enhanced Sensitivity to Noradrenaline of the Ainu. *International Journal of Biometeor.*, 1970, 14, 195-200.

Jensen, Arthur R., A Theoretical Note on Sex Linkage and Race Differences in Spatial Visualization Ability. *Behavior Genetics*, 1975, 5, 151-164.

Kantrowitz, Arthur, Controlling Technology Democratically. *The American Scientist*, 1975, 63, 505-509.

Kidd, K.K. and L.L. Cavalli-Sforza, An Analysis of the Genetics of Schizophrenia. *Social Biology*, 1973, 20, 254-265.

Kilbride, Janet E., Michael C. Robbins and Philip L. Kilbride, The Comparative Motor Development of Baganda, American White and Black Infants. *American Anthropologist*, 1970, 72, 1422-1428.

Klein, Thomas W. and John C. DeFries, Racial and Cultural Difference in Sensitivity to Flickering Light. *Social Biology*, 1973, 20, 212-217.

Klineberg, Otto, Press Release. October 16, 1956.

Kuo, Zing Yang, The Dynamics of Behavior Development. New York, Random House, 1967.

Lalouel, J.M. and N.E. Morton, Bioassay of Kinship in a South American Indian Population. *American Journal of Human Genetics*, 1973, 25, 62-73.

Lehrman, Daniel S., Gonadal Hormones and Parental Behavior in Birds and Infrahuman Mammals. In *Sex and Internal Secretion*, William C. Young (Ed.); William & Wilkins, Baltimore, 1961.

Lehrman, Daniel S., Semantic and Conceptual Issues in the Nature-Nurture

Problem. In *Development and Evolution of Behavior*. L.R. Aronson, E. Tobach, D.S. Lehrman and J.S. Rosenblatt (Eds.) W.H. Freeman, San Francisco, 1970.

Lewontin, Richard C., The Apportionment of Human Diversity. In T. Dobzhansky and Max Hecht (Eds.), *Evolutionary Biology*. New York, Appleton-Century-Crofts, 1970.

Lewontin, Richard C., The Genetic Basis of Evolutionary Change. New York, Columbia University Press, 1974.

MacLean, Charles J., Morton S. Adams, Webster C. Leyshon, Peter L. Workman, T. Edward Reed, Henry Gershowitz, and Lowell R. Weitkamp, Genetic Studies on Hybrid Populations. III. Blood Pressure in an American Black Population. *American Journal of Human Genetics*, 1974, 26, 614-626.

Maeroff, Gene I., Program to Spur College Hiring of Women and Minority Teachers Lags amid Continuing Controversy. The New York *Times*, December 28, 1975, p. 23.

Malina, Robert M., Skeletal Maturation Studied Longitudinally over One Year in American Whites and Negroes Six through Thirteen Years of Age. *Human Biology*, 1970, 42, 377-390.

Mead, Margaret, Theodosius Dobzhansky, Ethel Tobach, and Robert E. Light, Science and the Concept of Race. Columbia University Press, New York, 1968.

Miele, Frank and R. Travis Osborne, Racial Differences in Heritability Ratios for Verbal Ability. *Homo*, 1973, 24, 35-39.

Murphy, Edmond A., The Normal, and the Perils of the Sylleptic Argument. *Perspectives in Biology and Medicine*, 1972, 15, 566-582.

National Science Foundation, Report to the House Committee on Science and Technology by the Controller General of the United States: Administration of the Science Education Project: "Man: A Course of Study" (MACOS).

Nature, Editorial, September 27, 1974, Vol. 251.

New York *Times*, U.S. Immigrants Tape-Record Grim Memories of Nazi Holocaust. May 11, 1976, p. 25.

Nichols, Paul L. and V. Elving Anderson, Intellectual Performance, Race, and Socioeconomic Status. *Social Biology*, 1973, 20, 367-374.

Oelsner, Lesley, Reverse Bias: A New Injustice. The New York *Times*, January 18, 1976.

Reed, T. Edward, Number of Gene Loci Required for Accurate Estimation of Ancestral Population Proportions in Individual Hyman Hybrids. *Nature*, 1973, 244, 575-576.

Saksena, Sudha S., A Quantitative Method of Morphological Assessment of Hybridization in the U.S. Negro-White Male Crania. *American Journal of Physical Anthropology*, 1974, 41, 269-278.

Schneirla, T.C., Selected Writings. L.R. Aronson, E. Tobach, D.S. Lehrman and J.S. Rosenblatt (Eds.). W. H. Freeman Press, San Francisco, 1971.

Schulman, Gary I., Race, Sex, and Violence: A Laboratory Test of the Sexual Threat of the Black Male Hypothesis. *American Journal of Science*, 1974, 79, 1260-1277.

Schwebel, Milton, The Inevitability of Ideology in Psychological Theory. *Int. J. Ment. Health*, 1975, 3, 4-26.

Shore, Milton F. and Fortune V. Mannino, Mental Health and Social Change: 50 Years of Orthopsychiatry. AMS Press, New York, 1975.

Sieghart, Paul, A Corporate Conscience for the Scientific Community?

Nature, 1972, 239, 15-18.

Skeat, Walter W., An Etymological Dictionary of the English Language. Oxford University Press, Cambridge, 1968; first edition 1879.

Teyler, Timothy J., William M. Baum, and Michael M. Patterson, Behavioral and Biological Issues in the Learning Paradigm. *Physiological Psychology*, 1975, 3, 65-72.

Thoma, A., Dermatoglyphics and the Origin of Races. *Journal of Human Evolution*, 1974, 3, 241-245.

Tobach, E., The Meaning of the Cryptanthroparion. In *Genetics, Environment and Behavior*. L. Ehrman, G. Omenn, and E. Caspari (Eds.). Academic Press, New York, 1972.

Tobach, E., Social Darwinism Rides Again. In *The Four Horsemen: Racism, Sexism, Militarism and Social-Darwinism*. E. Tobach, Ed. Behavioral Publications, New York, 1973.

Trivers, Robert L., The Evolution of Reciprocal Altruism. *Quarterly Review of Biology*, 1971, 46, 35-57.

United Nations Document ST/TAO/HR/44: Seminar on the Dangers of a Recrudescence of Intolerance in All Its Forms and the Search for Ways of Preventing and Combatting It.

Valenstein, Elliot S., Persistent Problems in the Physical Control of the Brain. Forty-fourth James Arthur Lecture on the Evolution of the Brain, 1974, American Museum of Natural History, 1975.

Vetter, Betty M., Women and Minority Scientists. *Science*, 1975, 189, 751.

Wallace, Bruce and A. Srb, Adaptation. Englewood Cliffs, Prentice-Hall, 1961.

Wallace, Bruce, Topics in Population Genetics. W.W. Norton, New York, 1968.

Wallach, Michael A., Tests Tell Us Little about Talent. *The American Scientist*, 1976, 54, 57-63.

Walsh, R.J., A Distinctive Pigment of the Skin in New Guinea Indigenes. *Annals of Human Genetics*, London, 1971, 34, 379-388.

Whitehouse, H.L.K., Editorial Comment: 'A' Level Social Biology. *Biology and Human Affairs*, 1973, 38, 1-5.

Wilford, John Noble, Scientists Debate the Concept of an Impartial 'Court.' The New York *Times*, February 19, 1976.

Willerman, Lee and Richard E. Stafford, Maternal Effects on Intellectual Functioning. *Behavior Genetics*, 1972, 2, 321-325.

Wilson, E.O., Sociobiology. Harvard University Press, Cambridge, 1975.

INDEX

Abolition Society of London, 138
academic freedom, 12, 20, 23-24, 26-27, 32-34, 39-40, 44, 49, 52, 57-58, 114
American Anthropological Association, 51, 121
American Association for the Advancement of Science, 29, 146, 153; Committee on Scientific Freedom and Responsibility, 154
American Association of University Professors, 4, 20, 24
American Civil Liberties Union, 4, 20, 24, 38, 41
American Council on Education, 152
American Jewish Committee, 142
American Naturalist, The, 146
American Orthopsychiatric Association, 154
American Psychiatric Association, 37
American Psychologist, 8, 9, 11, 17, 18, 20, 22, 38, 49, 152
American Psychology Association, 3, 9, 11, 12, 16, 17, 18, 21, 22, 148, 153, 154; Council of Representatives, 150; Board of Social and Ethical Responsibility in Psychology, 152; Board of Scientific Affairs, 12, 152
Animal Behavior, 146
Archives of Internal Medicine, 103, 104, 105
Assimilation in American Life (Gordon), 84
Association of American Medical Colleges, 146
Atlantic Monthly, The, 59, 129

behavior, animal, 53, 63, 147, 148; human, 3, 4, 8, 12, 15-16, 19, 24, 52-54, 56, 62, 65, 67, 69, 73-76, 78, 82, 86-87, 91, 93, 95-96, 99, 129-30, 138-39, 142-51, 153
behavioralism, 19-20, 25, 76-79, 81, 88, 103, 124-25
"Behavioral Science and Society" (Layzer), 135
Behavior Genetics, 146

INDEX

Behavior Genetics Association, 51
Beyond the Melting Pot (Moynihan and Glazer), 85
biological warfare, 34
Biology and Human Affairs, 150
blacks, 18, 27-28, 41, 51, 55, 57, 59, 61-62, 65, 124-31, 138, 143, 155
Boaz, Franz, 154
Brown vs. Ferguson, 18, 124, 125
Bruno, Giordano, 115
Bulletin of Psychonomic Science, 146
Bureau of Science of the New York City Board of Education, 139
Burt, Cyril, 68, 128, 136
busing, 125

Cancro, Robert, 23, 24, 25; (cont.), 49-58
Carnegie Foundation for the Advancement of Teaching, 108
Chase, Allan, 14; (cont.), 99-112; 137
Chemical and Engineering News, 152
Child Welfare Research Station (University of Iowa), 129
City University of New York, 26
Civil Rights Movement, 26
Clark, Kenneth, 18
Committee on Social and Ethical Responsibility, 12
communists, 40
Congressional Committee on Science and Technology, 148
Crick, Francis H.C., 29
Cromwell, Oliver, 123
curricula, 4, 19, 26, 139-40, 147-50
Cutter Lecture in Biology, 102

Darwin, Charles, 3, 115
Davenport, Charles Benedict, 103-06, 108-10
Davis, Bernard D. 146, 147
Depression (American), 17
Derham, James, 138
Descartes, 52-53
desegregation, 18, 125-26, 143

Deutsch, Cynthia, 21
Deutsch, Martin, 18, 21, 28
D.M.Q. (developmental motor quotient), 129
Drucker, Ernest (cont.) 113-122
Dunn, L.C., 139

Edel, Abraham (cont.) 31-48; 137, 140, 150
education, 18, 44, 56-57, 64, 66, 77-78, 86-87, 89, 91, 93-94, 96, 107, 117, 126, 129, 138-39, 143, 146, 151, 154; compensatory, 71, 128, 132
Eichmann, A., 142
Einstein, Albert, 3, 39, 70
environment, 13, 14, 16, 20-21, 23, 35, 45, 53-56, 59, 61-64, 66-67, 69, 73-74, 82-83, 86-89, 91, 95-96, 101, 113, 118, 126-28, 130, 134, 138-39, 144-45, 152
environmentalism, 4, 8, 19, 26, 28-29, 49, 56, 123
"Equality of Educational Opportunity" (Coleman), 126, 128
Eugenical News, 99
eugenics movement, 51, 99, 103, 111
Eugenics Record Office, 103
Exploring Human Nature, 148
Eysenck, Hans J., 24, 61, 62, 70, 134

Fairfield, H., 24
Federation of American Scientists, 141
Feynman, Richard, 77
first amendment, 32
Freudianism, 27

Galileo, 3, 39, 70, 115
Garrison, P.E., 101-105
Galton, Francis, 128
genetics, 15, 28, 35, 43, 53-56, 60-67, 69-70, 74, 77, 103, 106, 113, 119, 126-31, 134-36, 144-45, 148-53, 155

INDEX

Genetics and Education (Jensen), 135
Gillispie, Charles C., 81-82
Glossary (Blount), 146
Goldberger, Joseph, 100, 102-06, 108-09

Harvard Educational Review, 18, 35, 59, 60, 126, 139, 153
Harvard Medical School, 146
hereditarianism, 4, 12, 14, 18-20, 23-24, 26-27, 29, 39, 44, 50, 60, 68, 71, 123, 128, 131, 143-44, 146-47, 149-50, 152-53
heredity, 3, 4, 8, 12-14, 16, 19-21, 23, 25, 28, 35, 38, 43, 45, 49, 54, 59-60, 99, 103, 107, 113-14, 121, 124, 126-29, 134, 138-39, 142, 144-45, 152
heritability, 60, 62-69, 72, 74, 126-28, 135-36, 142-43
Herrnstein, Richard J., 59, 60, 61, 63, 64, 65, 66, 67, 68, 70, 113, 114, 115, 129, 130, 131, 134, 135
High Cost of Thinking the Unthinkable," "The (Rice), 152
Hirsch, Jerry, 153
Hitler, Adolf, 3, 41, 44
Holmes, Oliver Wendell, 49
homosexuality, 37, 42
hookworm disease, 100-01, 103, 106
Hopkins Medical School, Johns, 108
Hullian learning theory, 27

Ingle, Dwight J., 153
inheritance, 4, 130, 149
"Innate and Learned Behavior," 147
intelligence, 13-16, 18, 28, 31, 33, 35-36, 38, 41, 43-45, 54, 60-62, 73-74, 121, 126-28, 134-35, 137-40, 143, 153
I.Q. (intelligence quotient), 18, 28, 29, 35, 36, 42, 51, 54-57, 59, 61, 63-72, 75, 77, 79, 96, 99, 113-14, 126-31, 134-35, 152

Jefferson, Thomas, 138
Jensen, Arthur, 13, 14, 18, 21, 23, 25, 28, 35, 36, 59-64, 66-68, 70-71, 75, 113-15, 126-31, 135, 139, 153, 154
Johnson, Lyndon B., 26, 41
Journal of Social Issues, 12

Kepler's laws, 77
King, James C. (cont.), 134-36

Layzer, David (cont.), 59-79; 135
lead poisoning, 137
Lewinian theory, 27
Life, 41
Lysenko, Trofim, 115

Malthus, Thomas, 110
"Man: A Course of Study" (MACOS), 147-48
McCarthy, Joseph, 40
Mendel, Gregor, 3, 39
meritocracy, hereditary, 60, 64-66
Milgrom, Harry, 139
Moynihan, Daniel, 41
Muncey, Elizabeth, 103, 105, 106

NAB/JOBS program, 87
National Academy of Sciences, 29
National Association fot the Study of Pellagra, 101
National Conference on Pellagra, 101
National Science Foundation, 147-49
natural selection, 74, 128, 142, 149
Nature, 119, 142, 146, 154
New England Journal of Medicine, The, 146
Newton, Issac, 70, 72, 76
New York Scientists Committee for Public Information (SCPI), 137, 139, 140
New York *Times*, 146, 151
New York *Times Magazine*, 59
Nixon, Richard, 41, 70
Norwood, Irving C., 100-01

NOVA, 146

Orientals, 55, 57

Page, Ellis B., 19, 49, 152
Padfield, Harlan (cont.), 80-98; 134, 135, 150
pellagra, 101-05, 110, 137
Pellagra Commission, The Robert M. Thompson, 101-07, 109
Pennsylvania Abolition Society, 138
Perspectives in Biology and Medicine, 153
Philadelphia Society for the Abolition of Slavery, 138
Piel, Gerard, 14; (cont.), 123-132; 141, 150
Popular Science Monthly, The, 100
Powledge, Tabitha, 119
Pritchett, Henry S., 107
Proceedings of the National Academy of Science, 146
Proshansky, Harold (cont.), 13-30; 143, 144
Psychologie, 153
psychology, 15-17, 19, 26, 75-76
Psychology Today, 152
public policy, 14, 31, 41-44, 52, 56, 70, 80-81, 84-89, 92, 94, 114-16, 118, 120, 124, 128, 132, 137, 140-42
race, 13-14, 25, 28, 31, 33, 35-36, 38-39, 41-45, 55, 62-63, 91, 114, 124, 127, 129, 132, 137, 139-40, 146, 153
"Racial Coalition," 24
racism, 51, 99, 124-25, 141-42
research, 15, 19, 31, 33-34, 69-71, 75, 79, 81, 114, 118; bias in, 82-83, responsibility for, 35-36, 40, 42-43, 45, 114-16, 119,-21, 132, 137, 141, 146, 153
Revolutionary War (U.S.), 138
Rickson, Roy E., 80-81
Rockefeller Foundation, 103

Rockefeller, John D., 101
Rockwell, George Lincoln, 58
Roman Catholicism, 40
Rosenthal, D., 24, 25
Rush, Benjamin, 138

Science, 119, 146, 151, 153
Science Curriculum Implementation Review Group of the Committee on Science and Technology, 147
"Science for the People," 24
Scientific Manpower Commission, 151
Shockley, William, 24, 26, 61, 62, 70, 113-15, 131
Sidel, Victor W. (cont.), 113-122
Siler, J.F., 101-05
Skinner, B.F., 76, 78
Smith Act, 32
Snow, John, 102
Social Darwinism, 128
Social Education, 147
Society for the Psychological Study of Social Issues (SPSSI), 12, 17, 18, 21, 23-25, 27-28, 40, 50, 152-53
SPSSI Commission on the Renewed Assault on Equality, 9, 11, 18-21, 50, 152-53
SPSSI Council, 11, 16, 18, 143-44, 152
socioeconomic status (SES), 59-61, 63-65, 67, 71, 77-78, 82-89, 91-96, 130
Stalin, Joseph, 3
Stevens, Joe B., 94
Stiles, Charles Wardell, 100-01
Stimson, A.M., 108
Students for a Democratic Society (SDS), 51
Sydenstricker, Edgar, 102

testing, 14, 35-36, 42, 56-57, 60-68, 73, 75, 99, 129, 150-51
Thorndike Award, 3
Tobach, Ethel (cont.), 142-158
tracking, 44, 126, 143

United Nations, 144
UNESCO, 143
United Nations Seminar on the Dangers of the Recrudescence of Intolerance, 144
U.S. Congress, 70
U.S. Constitution, 125
U.S. Office of Civil Rights, 151
U.S. Public Health Serviec, 99-100, 102, 104, 106
U.S. Supreme Court, 18, 99, 124, 143, 151
University Centers for Rational Alternatives, 4
University of Connecticut, 23-25

University of Michigan, 23

Vietnam War, 41

Western Rural Development Center, 92
whites, 18, 28, 51, 55, 61-63, 65, 99, 101, 103, 125-26, 138, 143
Williams, Curtis, 22
Wisdom, John Minor, 125
women, 27, 36
World War I, 14, 17-18, 32
World War II, 41, 119

Zeaman, David, 24, 25

DATE DUE

NOV 19 1997	
DEC 22 1997	
DEC 28 2001	NOV 01 1996

DEMCO, INC. 38-2931